Sarah T Stewart

John Delmonte
Dec-16, 1981

At the Crossroads of the Earth and the Sky
An Andean Cosmology

Latin American Monographs
Institute of Latin American Studies
The University of Texas at Austin

At the Crossroads of the Earth and the Sky

An Andean Cosmology

By Gary Urton

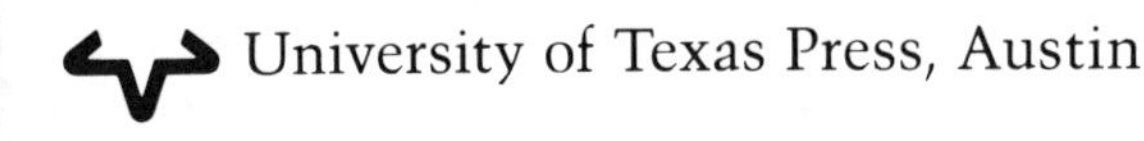
University of Texas Press, Austin

Library of Congress Cataloging in Publication Data

Urton, Gary 1946-
At the crossroads of the earth and the sky.

(Latin American monographs)
Bibliography: p.
Includes index.
1. Quechua Indians–Astronomy. 2. Quechua Indians–Philosophy.
3. Indians of South America–Peru–Cuzco (Dept.)–Astronomy.
4. Indians of South America–Peru–Cuzco (Dept.)–Philosophy.
5. Cuzco (Peru: Dept.)–Social life and customs. I. Title. II. Series:
Latin American monographs (University of Texas at Austin. Institute of Latin American Studies)
F2230.2.K4U77 1981 520′.985′37 81-4331
ISBN 0-292-70349-X AACR2

Printed in the United States of America

First Edition, 1981

For
my grandfather, R. D. Anderson
and
my father, Jason G. Urton

Contents

Foreword by R. T. Zuidema xiii

Preface xvii

Introduction 3

1. An Ethnographic and Calendrical Description of Misminay 17
2. The Organization and Structure of Space 37
3. The Sun and the Moon 67
4. Meteorological Lore 87
5. The Stars and Constellations 95
6. Collca: The Celestial Storehouse 113
7. Crosses in the Astronomy and Cosmology of Misminay 129
8. The Stars of the Twilight 151
9. Yana Phuyu: The Dark Cloud Animal Constellations 169
10. Summary and Conclusions 193

Appendix: Stellar Constellations Reported by Other Cuzco-Area Ethnographers 205

Notes 211

Bibliography 217

Index 233

Illustrations

Maps

1. The Region of Cuzco/Misminay 14
2. Misminay 39

Figures

1. Duties of the Months 24
2. Period of Frost and the Cycle of Maize 28
3. Saints' Days and the Planting of Maize 28
4. The Seasonal Cycle 29
5. Wooden *Pututu* 31
6. *Kay Pacha* and *Otra Nación* 38
7. The Center Path 41
8. Footpaths and the Quadripartition of Misminay 41
9. Flow of Water in Misminay 43
10. Footpath/Irrigation Canal Axes 44
11. Coordinate System in Juncal 45
12. House Clusters in Misminay 46
13. Two Views of Hierarchy 47
14. Azimuth Readings from Crucero 49
15. Horizon Profile from Crucero 50
16. Framework of Terrestrial Coordinates in Misminay 52
17. Top View of River/*Acequia* System 56
18. Side View of River/*Acequia* System 57
19. Alternating Axes of Milky Way from Fixed Point on Earth 58
20. Cosmic Circulation of Water via the Milky Way 60
21. Twelve-Hour and Twenty-Four-Hour Orientations of Milky Way in the Zenith 61

22. Solstices and the Seasonal Axes of the Milky Way 62
23. Cosmology of Misminay 63
24. Sections of the Sun 72
25. Sections of the Sun and Agricultural Cycle 73
26. Calendrical Periods of the Sections of the Sun 74
27. Center Sun from Misminay 75
28. Saints' Days and Center/Planting Sun 78
29. Lunar Synodic Terminology 82
30. *Pura/Wañu* Lunar Sequence 83
31. Lunar Terminology in Ocongate 84
32. Star-to-Star Constellations 108
33. Dark Cloud Constellations 110
34. Collca Axis 115
35. Collca and Atoq Axes 117
36. Tail of Scorpio and Full Moon 123
37. Planting Moons and Tail of Scorpio 124
38. Scorpio–The Celestial Plow 126
39. Cosmological Drawing of Pachacuti Yamqui 133
40. Cross of Passion 134
41. Celestial Contor 136
42. Directional Cross 137
43. Cross of Calvary 137
44. Stellar/Human Bridge in Misminay 138
45. Celestial Bridges in Sonqo 138
46. House Beams in Misminay 139
47. The Cross in the Preparation of *Chicha* 140
48. The Latin Cross 141
49. The *T*-Shaped Cross 141
50. Paired Crosses and the Solstices 142
51. Crosses of the Quarters 144
52. Paired and *Suyu* Cruciforms 145
53. *Suyu* Crosses and the Solstices 146
54. Crosses, Solstices, and the Milky Way 148
55. Terrestrial and Celestial Cruciforms 149
56. Twilight in the Arcadian Cycle 152

57. Periods of the Day and Night 153
58. Revolution of Dawn and Dusk 154
59. Twilight and the Lunar Phases 157
60. Orientations of the Stars of the Four Quarters 160
61. Quarters of the Milky Way 162
62. Limits of the Milky Way 163
63. Horizon Azimuths of Centers of Quarters of Milky Way 164
64. A Second View of the Cosmology of Misminay 165
65. *Yana Phuyu* Viewed from Misminay 171
66. Calendar of Toads and Agriculture 182
67. Calendar of Tinamous and Agricultural Guard Duties 186
68. Axis of the Fox 190
69. Cosmological Drawing of Pachacuti Yamqui 203
70. Irrigation-Canal Cleaning Altar in Tomanga 204

Tables

1. Potato-Field Names 33
2. Integration of the Potato and Wheat Cycles 34
3. Sacred Mountains of the Four Quarters 53
4. Topography and Symbolism of the Four Quarters 53
5. Horizon Calendar 73
6. Azimuths of the Center Sun 75
7. Catalogue of Quechua Stars and Constellations 96
8. Cycles of Maize and the Pleiades 121
9. Associations of the Celestial Storehouses 127
10. Forms and Meanings of Quechua Crosses (*Chaka*) 130
11. Celestial Crosses 135
12. Celestial Crosses and the Solstices 143
13. Twilight and Zenith Stars 155
14. Comparison of Lunar and Midnight Zenith Star Terminology 158
15. Heliacal Rise and Set Dates of the "Quarters" of the Milky Way 161

Foreword

The time to talk of stars is at night, which is not when most anthropologists do their fieldwork. This became clear to me one long, cold night when I walked with a guide and a Peruvian student in the Andean mountains. The guide and my student spoke of constellations and their relation to weather and crops—data to which the early Spanish chroniclers had been sensitive in the sixteenth century. Since then experts on Inca and Andean culture have repeated this scholarly information automatically, not aware that all of it could still be heard in any Peruvian village. I became aware that this general knowledge of stars derived from a living Andean tradition and not from textbooks on astronomy.

In 1973 Gary Urton and I initiated a project of studying Andean astronomy in the area of Cuzco, the old capital of the Incas, now a splendid colonial city and the bustling center of an area in southern Peru where a vigorous Andean culture imposes its image on modern life. Our aim was to reconstruct the old Inca astronomical system in Cuzco and to discover its importance to political, economic, and ritual organization in the imperial bureaucracy. Urton turned his interest to the astronomy and cosmology that structure the organization of a present-day Andean social system. Influences from Spanish culture and from the Catholic church touched the village of Misminay like any other normal group in the Andes. In Inca astronomy certain star-to-star constellations that cross the Milky Way were called *Chacana,* "cross-beam, ladder." Now we find names such as *Calvario,* "the cross of Calvary," or *Hatun cruz* (*Hatun,* "large" in Quechua), names that show the influence of the Christian idea of the Cross. But the constellations themselves still play the same role in defining the Milky Way as the principal organizing unit in the Andean astronomical-cosmological system.

The Milky Way, used to mark seasonal differences according to its changing positions in the sky, is central in the cosmology Urton describes. A somewhat similar orientation exists in "folk-astronomy" of Western Europe (not yet treated theoretically with the sophistication Urton applies to the Andean case). A cosmological outlook compatible in at least this respect may have facilitated a mutual understanding between Andean peoples and Spanish soldiers of nonurban origin, helping in the formation of an astronomical tradition as it exists today.

Urton has not fallen in the trap of too easy, "expert" opinions about similarities between his field data and those of the Incas or other cultures in South America. A wide reading has given him, however, a sensitivity that allows him to describe the local astronomical-cosmological ideas in all their ramifications and to a degree that does justice to a self-confident socioeconomic system and a living imagination. His careful recording of these ideas has become the motivation for a general ethnography.

Modern South American studies on mythology and religion, on kinship and social organization, now contribute significantly to general anthropological theory. Urton gives us examples of how new Andean issues can have a place in it. His treatment of the "organization and structure of space"—terrestrial and celestial—is a good case. The principle of quadripartition in cosmos and political organization is known of various American cultures. He demonstrates how it operates in practice in a society that does not need a formal division into four groups. Direct connections are established between spatial organization on land and in the sky, the horizon being the line of contact. The cross formed by the extreme positions of the Milky Way is used to relate the sky to the terrestrial quadripartitions formed by roads, irrigation canals, and directions toward distant towns. Urton offers us a dynamic analysis of data with their constant but changing aspects of importance for everyday decision making. He wholly revises the static perception of them as petrified survivals.

With a theoretically well-defined problem of a particular case before us, it becomes of interest to compare it, in its systematic aspects, to the astronomical system of Incaic Cuzco, geographically nearby and with a similar ecological setting. In Misminay, a small village, we find a highly complex system that integrates observed cycles of the sun, moon, and stars with agricultural cycles of planting and harvesting, of crop cycling, and of cycles pertaining to wild animal and plant life. Especially in terms of animal cycles Urton has pushed his later research further, demonstrating for which calendrical reasons certain animal names of "dark cloud" constellations are chosen. Despite the intricacies of the system in mnemotechnic terms, it serves the local needs of a village and does not have to be recorded by the people themselves in a formal way. In the case of Cuzco, the political-administrative needs of the empire made it necessary to register local calendars on quipus, knotted cords. By selecting comparable data, quantifying dates, and noting volumes of produce, Inca administrators recorded situations much like that in modern Misminay. But their records were of more importance to the administration than to the local people themselves, who relied on more subtle correspondences between cycles, rhythms, and rituals. Having analyzed the valley of Cuzco for some years now, starting originally with these administrative data, I am struck by the extent to which life in places like Misminay corresponds to the picture of local reality long ago.

In comparative study of Andean culture, local astronomies have assumed a critical value for analysis equal to that of local kinship and *ayllu* systems, political organizations as expressed in the "cargo" or *varayoq* systems, and economic systems. It is hoped that other studies will follow this model and guide in order to define Andean astronomical systems and Andean social and cultural variations in general.

R. Tom Zuidema
University of Illinois

Preface

My initial interest in Andean astronomy dates from the summer of 1973, when I traveled to Cuzco as a field assistant to R. T. Zuidema. Under a grant from the National Science Foundation, Dr. Zuidema was continuing his study of the ceque system of Cuzco. Our fieldwork involved locating and mapping the holy sites (*huacas*) of the ceque system which today have the same name as in Inca times. Dr. Zuidema was also interested in investigating the possible use of the ceques ("lines") as sightlines in the Incaic astronomical and calendrical system. My interest in Andean astronomy expanded as I found that a number of ethnographers working in the Cuzco area had recorded the contemporary existence of constellation names and astronomical beliefs similar to those reported over four hundred years ago in the Spanish chronicles. I decided to focus my work in the field on the collection of ethnoastronomical materials.

Between August 1975 and February 1976 I spent a month and a half camped at the archaeological ruins of Moray in the plain of Maras (Urubamba Province). I was studying with John Earls and Irene Silverblatt, anthropologists from the University of Michigan. Our research could be characterized as a climatological-archaeological-cosmological study of the ruins of Moray (see Earls 1976 and Earls and Silverblatt 1978 and n.d.).

Moray consists of four large "bowls" located in a much larger bowl, or geologic depression, all of which are the result of the slow seepage of water through the highly calcitic (calcium sulfate and carbonate) soils of the plain of Maras. The name of the community under study, Misminay, is related to this geologic process (mismiy, "the slow penetration of a liquid through a solid"). The largest of the four bowls of Moray, called Hatun Muyu ("great circle"), measures about 150 meters deep at the center by 150 meters across at the rim. Inside the four bowls, pre-Incaic and Incaic peoples constructed a series of concentric terraces. Ceramics excavated from three locations at Moray show an initial association with the Chanapata wares of the Cuzco Valley (R. Sananez Argumedo, personal communication, 1976). The presence of Chanapata ceramics suggests an initial occupation of the area of Moray by the first century A.D. (Lumbreras 1969). Classic Inca sherds are also commonly found on the surface of the site.

Hatun Muyu consists of twelve terraces. The masonry of the lower eight terraces is typified by reddish, roughly-worked stone and is similar in appearance and construction technique to the pre-Incaic site of Pikillaqta, located at the southeastern end of the Valley of Cuzco. The masonry of the upper four terraces is of a harder, grayish stone; also, the stones in the upper section are slightly better worked and fitted than those of the lower section. It is my impression from these observations that the lower eight terraces may have been constructed by pre-Incaic peoples (perhaps by the same population which occupied Pikillaqta), and that the upper terraces represent an expansion of the site during Inca times. The theory of Earls, and that of Patricio Arroyo Medina (1974), is that Moray was used by pre-Incaic and Incaic populations as a center for agricultural experimentation.

While carrying out our research at Moray, which consisted of measuring the soil temperature, air temperature, humidity, and so forth of different terraces at different times of the day and year, we gradually became involved with the women and children from the nearby communities of Misminay and Mistirakai who grazed their sheep and cattle on the thick grass which covers the terraces of Moray. Our time became more and more taken up with these visits, and it became obvious that we were of great interest to the people in Misminay and Mistirakai.

It was also apparent that the ruins of Moray are important not only in the economic (pastoral) life of the two communities, but in their religious life as well. The people told us fantastic stories concerning Moray. One very old man said that, during Inca times, the terrace walls were covered with plates of solid gold and silver in order to trap and reflect the light of the sun, the moon, and the stars; as a result, he said, it was never dark in Moray (Irene Silverblatt, personal communication, 1976). A man in Mistirakai told me that the eleventh Inca king, Huayna Capac, is now living underground at the center of the largest bowl of Moray. On two of the terraces there are large conical-shaped gypsum outcrops; one woman said that these outcrops are *ñust'as* ("princesses").

As in all Andean communities, reciprocity demanded that we visit our friends from the two villages. My initial introduction into the two communities, then, was a matter of accepting the responsibilities and obligations of reciprocity.

When I returned to Peru in August 1976 to begin my ethnographic fieldwork, it was perhaps a natural decision to look to Mistirakai or Misminay as a possible site. At this time, John Earls and Irene Silverblatt were still involved in their research at the site of Moray. While John was collecting and analyzing the climatological data (Earls 1976 and 1979), Irene had begun a study of the ethnohistoric documents in the Cuzco archives relating to the general area of Moray and Maras (Silverblatt 1976). It was decided to undertake an informal, three-part study centering on Moray: John would concentrate on the climatological and archaeological data from the ruins; Irene would continue her ethnohistorical research on the area; and I would work on the collection of ethnographic and ethnoastronomical data from the larger of the two communities, Misminay. In this way, our three separate studies would combine to give a comprehensive view of a single site in the Cuzco area from the earliest archaeological record of occupation down to the present day.

In mid-October 1976, I presented myself before Don Raymondo Juárez, *presidente* of Mullaca-Misminay, and requested that I be allowed to live for a year in the community. After following Don Raymondo around for a couple of days, during which time he gracefully tried to suggest that I would perhaps be more comfortable in some *other* community, he finally found me shelter in the unused storeroom of his father, Don Vicente Juárez, who lives in Mistirakai. Although my original intention was to live in Misminay, much larger and more centrally located with respect to the incorporated community of Mullaca-Misminay, for the first month of my fieldwork I was forced to divide my time between Mistirakai and Misminay. It soon became apparent, however, that I was accomplishing very little towards being accepted by either community. Finally, in November, I was offered the use of a storeroom in Misminay, next to the house of Benedicto Quispe Saloma and his wife, Manuela. It is of the time spent with Benedicto, Manuela, and their one-year-old daughter, Lucía, that I have the fondest memories of my fieldwork—the hours spent staring at Lucía as she breathed speechlessly into my tape recorder, the days spent planting and harvesting the crop, and the many evenings which they spent with me out of a genuine concern that I not be lonely. I extend to the three of them my deepest gratitude.

In addition to the friendship and guidance I have enjoyed from many people in the communities of Misminay and Mistirakai, I have also been fortunate to receive the advice, support, and kindness of many friends and colleagues during both the fieldwork and writing stages of this study. This book is an extensive revision of my doctoral dissertation, the initial fieldwork for which was undertaken from July 1975 to March 1976 and from August 1976 to July 1977. This revision has had the advantage of two subsequent visits to Misminay and other communities within the Department of Cuzco (July 1979 and August 1980) and from the critical reading of the manuscript by several colleagues in the field.

I would like first to acknowledge a substantial debt of gratitude which I owe to R. T. Zuidema of the University of Illinois (Urbana-Champaign), who was the head of my Ph. D. committee. Throughout my academic and personal acquaintance with him since 1971, Dr. Zuidema has served as my mentor in the most traditional, Old World form of that relationship. His support and guidance appear to me, in retrospect, to have been inexhaustible.

This study has also profited greatly from the initial, critical reading by the other members of my doctoral committee: Donald W. Lathrap, F. K. Lehmann, and Norman E. Whitten, Jr., of the University of Illinois Department of Anthropology; James Kahler of the University of Illinois Department of Astronomy; and Anthony F. Aveni of the Department of Physics and Astronomy, Colgate University. A number of others have read portions of the manuscript and have provided both criticism and encouragement: Billie Jean Isbell, Catherine Allen, Deborah Poole, John Earls, Irene Silverblatt, Bruce Mannheim, John A. Lawrence, Robert Harberts, Gerardo Reichel-Dolmatoff, Robert Goodwin, Steve Fabian, and John V. Murra. Thanks are also due to Don Wilson and Peter Dunham for their help with the art work, Annie G. Hurlbut for her support in Cuzco, and Lou Zeldis for help in Yucay. Finally, I express my deepest gratitude to Julia Meyerson, my wife, who has drawn most of the figures, read and criticized the manuscript innumerable times, and more recently has been a good companion in the field.

My fieldwork was made possible by financial support from the University of Illinois Department of Anthropology, the Wenner-Gren Foundation for Anthropological Research, the Organization of American States, the Sigma Xi Scientific Research Society, and the Research Council of Colgate University. Their assistance is most gratefully acknowledged.

At the Crossroads of the Earth and the Sky

An Andean Cosmology

Introduction

The almost complete absence of astronomical data in ethnographies of the Quechua-speaking Indians of the Andes encourages the belief that the study of ethnoastronomy[1] is something less than a serious pursuit; that it is, at best, an esoteric search for irrelevant or vaguely remembered scraps of lore which are of interest only to a few anthropologists and a handful of old people in each community. However, late one afternoon while doing fieldwork in the community of Misminay, Peru, a thirteen-year-old boy with whom I was pasturing sheep volunteered the information that, in preparation for planting, everyone in the community was watching the stars called Collca ("storehouse") very closely each night. I was aware that the early part of the rainy season had been extremely dry that year (1976) and only a few people had been adventurous enough to plant their crop of potatoes. There was much discussion about how desperate the situation would be if planting did not begin soon. Following up on the young man's statement, I asked why everyone was watching Collca. His answer, accompanied by a sharp glance, was simple: *porque queremos vivir* ("because we want to live").

It was clear from this one comment that the stars are discussed so seldom not because they are unimportant or uninteresting, but because their role in the life of the community is a matter of such great seriousness. The success of the crops, and with it the survival of the community, depends on the proper interpretation not just of a few apparent clues such as the rainfall, temperature, and wind patterns, but also of the more subtle messages provided daily and nightly by the celestial bodies.

Lack of interest in the astronomical knowledge of contemporary Andean Quechua-speakers is all the more obvious when compared to the relatively large body of literature on Incaic astronomy and cosmology. The Spanish chronicles and documents from the mid-sixteenth century on contain numerous references to the use of astronomical phenomena in Incaic science and symbolism. Several early Spanish chroniclers, such as José de Acosta and Bernabé Cobo, had a sophisticated knowledge of the Renaissance tradition of astronomy and astrology, and their interest resulted in a large corpus of descriptive materials on Incaic astronomy, which can be used in deciphering the role of astronomy in the sociocultural organization of pre-Columbian Andean civilizations.

Despite this wealth of material, relatively little progress has been made in this field. The most important difficulty is the absence of written records prior to the time of the Conquest. Because the Incas did not have a system of writing, or at least not one which has been deciphered (see Jara 1975 and Radicati di Primeglio 1965), we are forced to rely on the descriptions of Incaic astronomy which have been filtered through the minds of the sixteenth-century Spanish chroniclers. Even those chroniclers who were Indian or part Indian (Pachacuti Yamqui, Felipe Guamán Poma de Ayala, and Garcilaso de la Vega) recorded their descriptions of indigenous society and culture in a nonindigenous, European language (Spanish). Therefore, the ethnohistoric documents must all be viewed as European *interpretations* of what was seen or heard.

A second problem is also an ethnohistorical concern. The Spanish chroniclers recorded a tremendous amount of astronomical material, but since most were members of the clergy one of their primary motivations was to investigate thoroughly the civilization which they had conquered in order to more thoroughly destroy it. The most obvious example of this interest is in the document written by Padre Pablo José de Arriaga (1920). He describes the idolatrous images (*huacas*) he encountered in a number of central Andean communities and ends each description with a statement of how he ordered the huacas to be totally destroyed. This "extirpation of idolatries" also took the form of religious proselytizing. Thus, it probably took the Indians only a very short time to reckon that there was absolutely no profit in paying public homage to a carved rock or a body of water or a star. (Nathan Wachtel [1971] has written a good account of the structural disintegration of Incaic society and culture during the early years of the Conquest.)

As a result of this process, the formalized, state-level apparatus for maintaining the pre-Conquest sociocultural organization was destroyed and the system of astronomy apparently "disappeared" shortly after the Conquest. Yet, parts of the system survived, either because they were syncretized with western European Catholic concepts or because they operated at a level which did not directly threaten the newly imposed state (e.g., the agricultural uses of the system of astronomy). The astronomy of the contemporary Quechua-speakers contains both elements which are the result of the process of syncretism and others which appear to be direct descendants of the system as first described in the chronicles.

The third reason for the slow progress in this area of research is that scholars have so often approached the subject with preconceived ideas about what constitutes a system of astronomy. This leads to problems as basic as mistakenly assuming which way is "up"; that is, in the southern hemisphere is "up" north as it is in the northern hemisphere—or is "up" south? It is often assumed, moreover, that the cardinal directions are essential for an orderly system of astronomy and cosmology. However, if there is no fixed pole star marking cardinal south, as is true in the south celestial hemisphere, could not an orderly system of orientation be developed on the basis of noncardinal directions? [2]

A related problem has especially hampered progress in Andean ethnoastronomy. It is the tendency to make assumptions about what other people *must* see when they look at the sky. Many late-nineteenth and early-twentieth-century scholars (especially G. V. Callegari, Jean Du Gourcq, and Stanbury Hagar) attempted to reconstruct the Incaic constellations as though they were identical to the Greek and Roman constellations. For example, Hagar made a careful study of the cosmological drawing of Pachacuti Yamqui (1950:226) and concluded by equating twelve figures in the drawing to the twelve constellations of the European zodiac (Hagar 1902:283). In a similar vein, Antonio Tejeíro (1955) formally related each of the constellations of the contemporary Aymara-speaking Indians of Bolivia to a constellation in the European tradition.

Now, the point of objection is that Gemini, or Capricorn, or Cancer have as much business floating over the imperial city of Cuzco as the "dark cloud" constellation of the Llama has floating over the sanctuary of Delphi. None of these constellations exists in the sky *unless* a particular culture agrees on its existence. The stellar constellations of western Europe and the Mediterranean are the constructs of a long cultural tradition which has its roots in Egypt and Sumeria. Without the cultural tradition, the constellations have no calendrical, symbolic, or cosmological meaning. In addition, it is often assumed by Western-trained scholars that there is some real relationship between the order of the stars and the shapes which the classical Western civilizations projected onto the celestial sphere. This, again, is not the case. Almost every culture seems to have recognized a few of the same celestial groupings (e.g., the tight cluster of the Pleiades, the *V* of the Hyades, the straight line of the belt of Orion), but the large constellation shapes of European astronomy and astrology simply are not universally recognized; the shapes were projected onto the stars because the shapes were important objects or characters in the Western religious, mythological, and calendrical tradition. Thus, it is wrong to assume that different sociocultural groups will project the same shapes onto the stars—or even that different cultures will have the same motivation for ordering the stars into constellations.

Despite the problems which have hindered the study of Incaic and contemporary Andean astronomy and cosmology, there have been important contributions in the field. The literature can be divided into the three subject categories with which most research in this area has dealt: (a) archaeoastronomical[3] studies on the calendar and the solar observations of the Incas; (b) ethnoastronomical studies on Incaic cosmology and astronomy; and (c) the ethnoastronomical study of the calendrics, cosmology, and astronomy of contemporary Andean Quechua-speakers.

Archaeoastronomical Studies. Several ethnohistorical accounts state that the Incas in Cuzco erected a number of pillars along the horizon for the purpose of observing the sunrise or the sunset or both at different times of the year. The solar pillars are said to have been used to divide the movement of the sun into periods roughly corresponding to monthly (i.e., lunar) time periods. They also are said to have been used to observe the movement of the sun along the horizon in order to determine the time for planting in the Cuzco Valley. Ethnohistoric descriptions of the solar pillars can be found in the chronicles of José de Acosta (1954, bk. 6, chap. 3); Juan diez de Betanzos (1924, chap. 15); Bernabé Cobo (1964, vol. 2:158); Garcilaso de la Vega (1966:116-117); Juan Polo de Ondegardo (1916:16); and Pedro Sarmiento de Gamboa (1942:93) and in *Discurso de la sucesión y gobierno de los Yngas* (Anonymous 1906:151).

Two major problems have plagued the attempt to reconstruct the methods used by the Incas to observe the sun by means of horizon pillars. First, owing to the systematic destruction of indigenous idols during the years following the Conquest, none of the solar pillars is extant. All attempts to understand the methods of solar observation therefore must be based on ethnohistoric accounts of the number and placement of the pillars. This leads to the second major problem, the apparent contradictions in the ethnohistoric accounts. In these accounts the number of pillars ranges from four to sixteen. In addition, some accounts suggest that the pillars were confined either to the eastern or to the western horizon, others say they were located in both the east and the west. Horacio H. Urteaga (1913) made an early attempt to analyze these contradictions, but his work was little more than a summary statement of the ethnohistoric problem. Rolf Müller (1929 and 1972), working in Cuzco in the 1920s and 1930s, also studied this problem and made some interesting suggestions for the astronomical, especially stellar, alignment of streets in Cuzco. Recently, R. T. Zuidema (1977a and 1978b) and Anthony F. Aveni (n.d.a) have systematically studied the chronicles and have spent three summers (1976-1978) in the Cuzco Valley taking archaeoastronomical measurements. They made suggestions that allow us to resolve the ethnohistorical contradictions in the numbers of pillars and their placement by relating the pillars to different methods of solar observations made for different purposes; thus, different pillars and different numbers of pillars would have been used for different observational purposes.

As stated above, it is commonly accepted by students of Inca culture history that solar observations were used by the Incas to fix the dates for sowing and as a basis for the calendar. However, once the references are made, it is assumed either that the setting up of a solar calendar is a fairly simple affair, in which case the Incas were obviously capable of it, or that it is a very complicated, theoretical affair, in which case the Incas may or may not have been capable of it. An example of the latter approach is John H. Rowe's conclusion that the Incas "probably made no use of solstices and equinoxes in their calendar, whether or not they displayed enough theoretical interest to observe them. Their months were lunar, and they seem to have had no very exact way of adjusting them to the solar year. Probably the count was arbitrarily adjusted when the annual solar observations indicated that it was seriously wrong. (1946:328)" Although Rowe (1979) has slightly modified his position with regard to the interest of the Incas in the solstices and equinoxes, the quotation serves to communicate a long-prevalent attitude regarding the relative complexity and sophistication of Incaic astronomy (see the replies to Rowe in Aveni, n.d.b; and Zuidema, n.d.b).

Without question the establishment and maintenance of a calendar system *is* a very complex matter, and certainly if a calendar was used to time something as essential as the sowing of the crops, one must conclude that the calendar system worked, and that it worked very well. Supposing that the adjustments were made "arbitrarily" (by the state?), as Rowe suggests, one must then assume that precise observations and calculations were being made and that accurate records were being kept. On the other hand, if the calendrical correlations are the result of nonarbitrary, social consensus, the problems of maintaining a calendar are even more complex. For example, in a study of a community in Ethiopia, David Turton and Clive Ruggles (1978) have analyzed an extremely complicated social process of lunar intercalation based on the community's agreement to *disagree* about which month it is, and Edmund R. Leach (1950) provides examples from Oceania in his discussion of the structure and operation of four calendar systems based on the principle of social or consensus intercalation. By whatever means, calendrical correlation is a difficult task and one which must be undertaken by every society; the alternative is economic, political, and psychological chaos. (We experience our own periodic, biological discomfort when merely changing one hour between standard time and daylight-saving time.) Having made such assumptions concerning the importance of the calendar system we must next try to understand how the Incas observed the sun, how the divisions of time based on solar observations were translated into an agricultural and ritual calendar, and how the lunar and solar periods were integrated.

In a number of articles beginning in 1966, R. T. Zuidema analyzed various aspects of the Inca calendar system. The foundation of Zuidema's research is his earlier (1964) work on the *ceque* system of Cuzco. The ceque ("line") system, as described in the chronicles of Polo de Ondegardo (1916:43) and Bernabé Cobo (1964, vol. 2:169-186), was the system whereby the Incas organized the actual topography of the Cuzco Valley in accordance with their social and religious organization. In the ceque system, there were

> approximately four hundred holy sites [huacas] in and around Cuzco, consisting of stones, fountains, or houses which . . . were of particular significance in Inca mythology or history. These sites were divided into groups. Every group of sites was conceived of as lying on an imaginary line, called a ceque; all these lines were deemed to converge in the centre of Cuzco. The maintenance and worship of the sites lying on these lines was assigned to certain social groups. (Zuidema 1964:1)

In Zuidema's first study of the Inca calendar system (1966), he analyzed the numerical properties of the huacas in the ceque system and proposed that, in addition to organizing the social and religious components of Inca society, the ceque system coincidentally served as the mechanism for organizing the calendar. Thus, the ceque system merged the ordering of space with the ordering of time. The fact that certain of the holy sites located on the ceques are mentioned in relation to the pillars (*sucancas*) used in the observation of the sun led Zuidema to propose that the ceques were used for more than numerical calculations: since the ceques are explicitly said to be straight lines (Cobo 1964, vol. 3:169) they served as sightlines for observing the rise and set of celestial bodies (Zuidema 1977a and 1978b especially). In this way, the ceque system would accomplish the reckoning of time by correlating terrestrial units of space with the rise and set of celestial bodies—in effect, merging terrestrial space with celestial space.

Because the Cuzco solar pillars no longer exist and the ethnohistoric accounts of them present several problems for reconstructing the observational methods in the field, Zuidema's approach provides a way of studying the problem within a larger social and religious context. In 1973 he and I spent two months in the field, recording and mapping toponyms in the Valley of Cuzco. We found that one-third to one-half of the names given in the chronicles for the huacas of the ceque system are still known in Cuzco. In view of the fact that the organization of huacas along the ceques is analogous to a *quipu* (the mnemonic device of knotted strings used by the Incas to record calendrical information and other numerical data), Zuidema's study of the ceque system affords us the best analytic approach for reconstructing the Inca calendar and the system of astronomy.

In the present study, I have not attempted to synthesize and analyze all of the Incaic and contemporary ethnoastronomical data, although astronomical data from my fieldwork are discussed in terms of selected ethnoastronomical data drawn from the Spanish chronicles. I have also made use of ethnohistoric materials when appropriate for analyzing or illustrating concepts for which my own data are incomplete.

Incaic Ethnoastronomical Studies. This category refers to the attempts which have been made to analyze (on the basis of the chronicles) the constellations of the Incas and the structure of the cosmological system. Many of these data directly relate to the more general problems just discussed.

The ethnohistoric documents provide several accounts of the Incaic constellations. The most extensive listings are found in the writings of Francisco de Avila (1966:chap. 29), Polo de Ondegardo (1916:305), Pachacuti Yamqui (1950: 226), and Diego González Holguín (1952); other, less-complete accounts are discussed where relevant in the body of this study.

As noted earlier, many early students of Inca culture history attempted to relate the constellations of the Incas to the constellations of the Western world, an approach that makes a number of unwarranted assumptions about the perception of form and of the relationship between a system of constellations and its specific sociocultural context. Two more general, conceptual problems are also involved. The first is the assumption that the principal constellations will always be oriented along the ecliptic (the path of the sun, the moon, and the planets among the stars). The twelve major constellations of Western astronomy and astrology are oriented along the plane of the ecliptic, as are the principal constellations of most other cultures in the northern hemisphere (see Kelly, 1960). It has been uncritically assumed, therefore, that the Incas utilized the same plane for their celestial orientations. As I have shown elsewhere (Urton 1978b), and will show more completely here, the principal plane of celestial orientation used by both the Incas and the contemporary Quechua-speakers is the Milky Way, not the ecliptic. This difference in orientational planes means that, with respect to the astronomical and cosmological orientations used in the northern hemisphere, there is a roughly 90° shift in the Incaic (Quechua) system of astronomical orientations. We will see in the course of this study why the shift was necessary and what its structural and orientational effects were on the cosmological system.

Beyond this problem of orientation, attempts to reconstruct the Incaic constellations have assumed that even if the Incas did not have the *same* constellations as the Western world, they at least must have had the *same type* of constellations. The constellations with which most Western scholars are familiar are "star-to-star" constellations, celestial shapes formed by conceptually linking together neighboring bright stars. A good example of this approach is seen in the work of Robert Lehmann-Nitsche (1928), who attempted to reconstruct the

Incaic constellations through analysis of the ethnohistoric documentation. Lehmann-Nitsche based his study on the cosmological drawing in the chronicle of Pachacuti Yamqui (1950:226), which shows a number of constellations and identifies them by name. In one especially interesting section, Lehmann-Nitsche discussed the testimony of the chronicles regarding a constellation referred to as a *llama*. He concluded that those chroniclers who describe the Llama as a "black spot" (*mancha negra*) in the Milky Way were mistaken, because Lehmann-Nitsche knew that constellations are made up of groups of stars. In dismissing Garcilaso's account, he says that "despite the usefulness of Garcilaso, he was misguided when he fixed [the Llama] in the spots of the Milky Way; his account ceases to be of interest" (Lehmann-Nitsche, 1928:36; my translation).

As we will find later, the "dark cloud" animal constellations are patches of interstellar dust which cut through the Milky Way; to omit these celestial forms as constellations is to omit at least half of the complexity, and the beauty, of the Incaic system of astronomy. A recent article on the dark-cloud constellation of the Llama analyzes its significance in the ritual and agricultural calendar of the Incas (Zuidema and Urton 1976). In an earlier study, Zuidema (1973) recognized the structural importance of the dichotomy between star-to-star and dark-cloud constellations when he demonstrated that the two types of constellations were symbolically related to different social groups (*ayllus*) in the community of Recuay during colonial times.

A number of other important contributions have been made in the area of Incaic astronomy and cosmology. In a study combining Zuidema's work on the ceque system with his own ethnographic materials from the area of the Río Pampas (Ayacucho), John Earls (1972) analyzed the spatial and temporal characteristics of the Andean cosmological system (see also Earls 1971 and 1973). He demonstrated that the structural properties of the Incaic organization of space and time (as seen through such institutions as dynastic succession and age classes, for example) are structurally analogous to the organization of activities and institutions in a contemporary community in the central Andes. Earls's work is especially important to the present study because he argued that the organization of the Valley of Cuzco via the ceque system is homologous with the structure and organization of the celestial sphere in Inca cosmology.[4]

More recently, John Earls and Irene Silverblatt (1978 and Earls 1979) have undertaken a systematic study of the ruins of Moray (Department of Cuzco) in order to analyze the correlation between Inca cosmology and the structure of an archaeological site which, they hypothesize, was used for agricultural experimentation. The community in which I carried out my fieldwork, Misminay, is located a ten-minute walk from Moray.

Ethnoastronomical Studies of the Contemporary Quechua-speaking Indians of the South Peruvian Andes. Aside from the present work, no other studies are devoted entirely to the analysis of the astronomy of the Quechua-speaking Indians of the Peruvian Andes.[5] However, a number of ethnographers have collected the names, and, less frequently, the celestial locations of various Quechua constellations. A list of some of the major collections appears in the Appendix; data from several of these are used extensively for comparative purposes.

One of the most useful listings of astronomical data is from a work by P. Jorge Lira entitled *Farmacopea tradicional indígena y prácticas rituales* (1946). Padre Lira's data are in a sense unique because he places the astronomical data in in a firm calendrical and cultural context; that is, he describes the various celestial phenomena used in the crop predictions of August and February. Celestial observations used in weather and crop predictions have also been reported by Jorge W. Bonett Yépez (1970), Bernard Mishkin (1940), and Juan V. Tuero Villa (1973). Additional collections of Quechua constellations and astronomical data to which I frequently refer are from Juvenal Casaverde Rojas (1970), Juan Núñez del Prado (1970), and Christopher Wallis (personal communication, 1976).

While in the field, I made extensive use of a star map drawn by Sergei Gaposchkin from 1956 to 1957 (Gaposchkin 1960:293). The map is the best available for ethnoastronomers working in the southern hemisphere because it is drawn with the south celestial pole in the center and because Gaposchkin so faithfully rendered the dark spots in the Milky Way. I have often thought that were it not for the diligence and patience of Dr. Gaposchkin in producing this map, it might have taken much longer for us to begin to understand the significance of the "dark cloud" constellations in Quechua and Incaic astronomy. I used a different copy of the map each time I discussed the stars with an informant, even at night when the stars were being identified in the sky. In this way, I was able to make a comparative study in order to determine which constellations were identified by which informants and exactly when the identifications were made (the name of the informant, the date, time, and phase of the moon were listed on each map).

The Choice of a Community for Fieldwork

The principal aims of my fieldwork were to investigate the extent to which contemporary astronomical beliefs might still operate as a coherent "system" (in conjunction with the calendar of agricultural duties, festival cycles, and so on), and to determine the extent to which the contemporary astronomical data might help in understanding Incaic astronomy, cosmology, and calendrics. With these goals in mind, it was apparent that a community for fieldwork had to be chosen according to a few rather specific criteria: first, the community should have some demonstrable (i.e., ethnohistorically recorded) relationship to Cuzco during early colonial times; second, it had to be within the general area of Cuzco in order to suppose that both the community and Cuzco had been subjected to approximately similar historical, organizational, linguistic, ethnic, and other influences; and finally, the climatic and ecological setting had to be almost identical to that of Cuzco.

The last criterion was perhaps the most critical because, if either the climate or ecology were very different between the two communities, it would be difficult to assume that the local calendar systems were similar. In other words, when a calendar system for use in a particular community or ecological zone is constructed, the calendar must meet certain minimal requirements if it is to be at all functional. The calendar should be fairly accurate in indicating periods ofraininess and dryness; it must be very accurate in predicting the beginning and end points of periods of frost; and it should contain a number of discrete cycles for use in timing such things as the planting of different crops and the periods of animal gestation and birth.

The altitude of Cuzco (3220 meters) places it near the upper limit for the cultivation of maize. In the mountains around Cuzco, wheat, potatoes, *oca, ullucu, quinua,* and other high-altitude crops are grown by utilizing a system of fallowing. Thus, the climate and altitude ranges in the Cuzco Valley demand a calendar system which can accommodate every cycle, from the long growing season of maize (which can be shortened somewhat by irrigation) to the shorter growing season of high-altitude tubers. In earlier fieldwork, I had had the opportunity to observe communities at opposite ends of this ecological and calendrical spectrum.

In 1975 and 1976, I spent a week and a half in the community of Sonqo, (district of Qolqepata), which is located high in the puna ("tundra") at an altitude of 3900 meters. Aside from the obvious fact that the puna offers few climatic attractions, I found Sonqo to be very different from Cuzco in general life style. The puna climate and land can sustain pastoralism and tuber cultivation, but it is too high and cold for the cultivation of maize. Therefore, it appeared very likely that the puna calendar and astronomy would have to serve different needs from the calendar and astronomy of Cuzco.

On the other hand, I had also spent a month and a half in the community of Yucay in late 1975. Yucay is located in the Vilcanota Valley, about three kilometers upriver from the provincial capital of Urubamba; it is situated at an altitude of 2865 meters (about 350 meters lower than Cuzco). The altitude of Yucay is suitable for the cultivation of maize, beans, wheat, and other low-altitude crops; however, the high-altitude tubers and grains consumed in Yucay come from communities higher up on either side of the Vilcanota River. Although Antoinette Fioravanti (1973) has shown that a complex pattern of interdependence exists between Yucay and certain puna communities (especially the village of San Juan), the situation is nonetheless that of two essentially different environmental and ecological zones integrated through *compadrazgo* (ritual co-parenthood), markets, and reciprocal labor (Isbell 1978, Murra 1972, and Brush 1977). An integration of communities and ecological zones through these kinds of formal economic and social ties is a different order form of organization than one in which the communities merge into a *single* system; the latter implies a single calendar with a single hierarchy of authority maintaining it.

On the basis of these prior field experiences, it was obvious that for the specific objectives of my fieldwork, I should live in a community which was neither above 3500 meters nor below 3000 meters. The land cultivated by the *campesinos* of Misminay is located between an altitude of 3200 and 3600 meters. Within this altitudinal range, the following crops can be grown, and are grown in Misminay: quinua, oca, ullucu, several varieties of potatoes, wheat, beans, peas, maize, and even cherries. Thus, Misminay met the criterion that the community should have a climatic and ecological setting almost identical to that of Cuzco. As for the requirement of its being relatively near the Valley of Cuzco, Misminay (province of Urubamba, district of Maras) is located about fifty kilometers northwest of Cuzco; and though one-third to one-half of the men in Misminay are bilingual (Quechua-Spanish), the Cuzco dialect of Quechua is spoken almost exclusively within the community.

Misminay also meets, indirectly at least, the criterion that the community should have an ethnohistorically recorded relationship to Cuzco. Misminay is one of four villages which in 1964 were united to form the incorporated community of Mullaca-Misminay (see map 1). The other two villages are Santa Ana and Mistirakai. I occasionally heard the four communities referred to as four different ayllus, but they are more commonly referred to as *comunidades* ("communities"). In a document dating to 1791 (A.N.L. Legajo XVIII cuaderno 456 año 1791), Mullaca is included within the Collana ayllu of the community of Maras. The total composition of Collana ayllu in the eighteenth century included the Yanacona, Collana, Oyola, Maras, and Mullaca (Zuidema 1964:99; also O'Phelan Godoy 1977:117). In Zuidema's study of the ayllus in the area of Maras and Anta (1964:178 n. 11), he further concluded that the Indians of Maras were used as policemen, messengers, and spies for the Inca. Finally, in a study of the archaeological site of Moray, Patricio Arroyo Medina (1974) states that the following groups were located in the area of Laguno Huaypo (see map 1) and the plain of Maras at a very early time: Pampakonas, Pauchis, Mismis, Pukaras, Pichinqotos, Lares, Qoripujios, and Chillkapampas.[6] The Mismis are said to have been situated near a mountain called Wañumarka (Arroyo 1974: 18-19). Misminay is located in a small valley directly south of Apu Wañumarka, and Wañumarka is the principal sacred mountain of the community. Thus, with

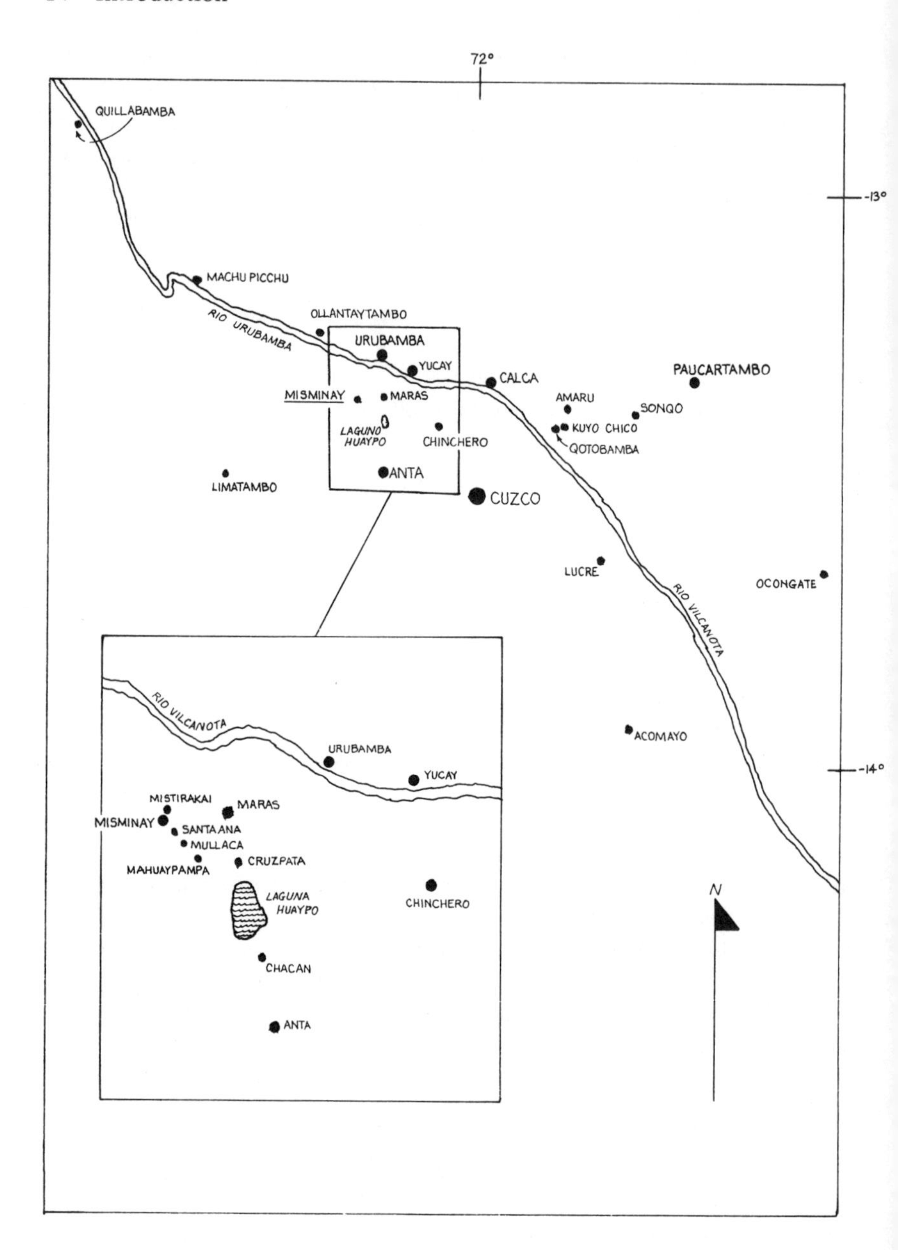

Map 1. The Region of Cuzco/Misminay

some certainty it can be concluded that the area of Misminay (i.e., Mullaca-Misminay and Moray) was occupied prior to the time of the Conquest and probably by the ancestors of the present inhabitants. An early occupation of this area is supported by the presence of pre-Incaic and Incaic ceramics at the site of Moray.

1. An Ethnographic and Calendrical Description of Misminay

In collecting the ethnoastronomical data it is important to understand their practical use as well as their more general cosmological significance. The presentation of an esoteric cosmological system adds nothing to our understanding of another culture; it becomes an artifact, an anthropological curio, which may or may not have a relationship to the lives of the people. As Professor John Murra once remarked, astronomy and cosmology are meaningless unless they are related to *la comida* ("food").

In the following ethnographic sketch of Misminay, the material is grouped according to the general time periods by which activities are organized. These time periods include the twenty-four-hour period of the day, the cycle of weeks, the activities of the months, the two seasonal divisions of the year, and multi-annual cycles. As we proceed, we will see how these calendrical periodicities and cycles of activity are related to different astronomical phenomena.

The Arcadian Cycle

If a "typical day" were ever to occur in Misminay (which it never did the entire time of my fieldwork), the following outline would be a reasonable description of it. The day begins around 4:00 A.M. when the village slowly comes to life through sound. As the sky lightens, one hears the repetitious whistle of the *pichiko*, a small gray and black bird with a red collar whose call is a sharp *"pichi, pichi."* To the call of the pichiko is soon added the crowing of roosters and the sound of flapping wings as hens and roosters emerge from their coops to forage for scraps of food thrown out of the houses during the night. The stage is set for a burro to begin its loud, persistent, and unpleasant braying; the day has begun.

Curiously, the greatest amount of noise in the village occurs at dawn and dusk; unnecessary noise, even that of animals, is not easily tolerated at any other time. If a burro begins to bray during the daytime, people become very upset and the wretched beast is subjected to every imaginable form of physical and verbal abuse. A burro loaded with a heavy burden is allowed to bray for only as long as it takes his owner to catch him up and plant a well-aimed kick in his side. Noise marks the transition from day to night and from night to day; it is inappropriate at other times. This is also true for the pichiko, which—although I have often heard its distinctive call during the daytime—is *said* to call only at dawn and dusk. Thus, the sound of the pichiko is "appropriate," or recognized, only at these transitional (twilight) times.

As the animal sounds slowly die down, the day's activities begin around 5:30 A.M. with a light snack of coffee (always heavily sugared) and popcorn or cold leftovers such as boiled corn, beans, and potatoes. Breakfast is then prepared by the women and eaten between 7:00 and 8:00 A.M. During the hour or so between the early morning snack and breakfast, men busy themselves with such household chores as collecting firewood and going for water or else they visit the houses of neighbors or relatives to organize the day's work. This is also the time when small community meetings take place near the central chapel called Crucero. At this time, the *teniente gobernador* (the community official who oversees most day-to-day, communal activities) may call a brief meeting to arrange communal work projects or to determine the day's pattern of irrigation. The larger community meetings, which take place every couple of weeks, usually consume at least half a day and are concerned with long-range planning and policy. The small, "pre-breakfast" meetings are concerned only with the specific plans for that day.

Breakfast usually consists of two or three servings of a soup made of potatoes, beans, cereal, and boiled corn and potatoes. It is occasionally more elaborate, such as when fried meat or potatoes are served. If available, bread, a delicacy purchased in Maras or Urubamba, is also eaten at breakfast.

When breakfast is over and the dishes and pots have been cleaned and stacked, the day's work begins; the division of labor by sex now becomes most evident. After wrapping up a lunch of boiled corn and beans, the women and small children begin to herd the cattle (usually only a few head) and sheep (anywhere from ten to forty) out of the corral toward the pastureland. Most pasturing is done in the mountains higher up around Misminay or lower down near the ruins of Moray. If a family has an especially large herd, or if a woman is responsible for pasturing the sheep of another family, the husband may join her in driving the livestock out to pasture in the morning and in bringing them back in the evening. This is almost always the case when driving bulls; men usually take the bulls to pasture and tie them to a short stake which is driven into the ground. It is also common for a man to pasture the cattle near where he is working in the field while his wife pastures the sheep elsewhere.

After the women have reached the pasture with the livestock, they pass the day with their attention equally divided by at least three simultaneous chores: caring for the baby (a permanent fixture on the backs of most women), spinning wool into yarn, and keeping an eye on the herd. A certain amount of socializing goes on throughout the day among the women as they move the herds slowly across a mountainside. One or two women may eat together at noon, but it would be unusual to see large groups of women visiting in the mountains while pasturing.

While the normal duties of women revolve around cooking and pastoral activities, men are primarily concerned with agriculture. The agricultural activities vary a great deal depending on the time of year. (The major duties of the agricultural cycle are described below in the section on the "months.") When working in the field, men usually take one work break in midmorning and another in midafternoon. If only a few men are working together, they bring small jugs of *chicha* (fermented corn beer) with them and drink it during these regular breaks and off and on during short rest stops (e.g., after plowing a few rows with a foot plow). The chicha jugs are usually empty by midafternoon, and it is common to see young boys flying down the hill to refill their father's chicha jug. Coca is also chewed during work breaks, although the use of coca in Misminay is much less frequent than in puna communities (see Bolton 1976; Gow 1976a; Isbell 1978; and Wagner 1976 and 1978).

Women also carry small jugs of chicha with them while pasturing the livestock. Coca, however, is almost never used by women; it is said to be too "hot" for them. A woman in Misminay once carefully explained to me that coca was an "invention" originally intended for men, not women. Therefore, she said, women do not chew it. As we talked, we were joined by an old woman who asked me for coca and after receiving a handful, promptly began chewing it. In this connection, it is also curious to note that, although the principal coca diviners (*paqos*—people who are paid to divine the future, locate stolen articles, and the like) in Misminay are men, the first two coca divinations I witnessed were performed by women.

Concerning the relation of the division of labor to daily activities, we can make the generalization that the male day is "agricultural time" whereas the female day is "pastoral time." For men, the passage of time (i.e., the difference between any two days) is marked by the cycling and duration of different agricultural duties and the growth of the crops; for women, time is marked by the alternation of pasturelands and the growth of the animals. These different modes of time are in turn related to a number of biological and astronomical cycles (the former are described later in this chapter and in chapter 9; the latter account for the balance of this study).

The main differences in the daily activities of children are determined by age and sex. Until the age of about five or six, all children spend the greater part of the day either pasturing the livestock with their mothers or playing around the house under the care of an older sister. Provided that a man is working at a leisurely pace and has time to care for a child, very young boys may be carried to the field in a *lliqlla* ("shoulder cloth") on their father's back. However, children of this age are generally not taken to the fields when the men are working in large work parties.

Beyond the age of five or six, the differences in the daily activities of children are determined by sex. Boys from six to nine years of age spend at least half a day in the school in Santa Ana (a ten-minute walk to the southeast); boys from nine to twelve years old walk an hour to an hour and a half in order to attend school in Maras. After school, the boys spend the afternoon playing or running errands for their parents. Girls certainly are not prohibited from attending school either in Santa Ana or Maras, but it is not very common for them to do so. Most young girls remain at home and gradually assume more responsibility in herding the sheep, caring for younger children, and spinning wool.

Thus, sex plays a major role in determining whether a child is exposed to the outside world (school) or is restricted to the community. Since the classes in the Santa Ana school are taught in both Quechua and Spanish, and those in Maras are taught mostly in Spanish, boys are slowly prepared for dealing with the outside, Spanish-speaking world of the mestizo. Very few of the girls or women in Misminay speak Spanish, a fact which reinforces the confinement of their activities to the village. Oscar Núñez del Prado (1973:45) reports a similar pattern of male-female matriculation in Kuyo Chico: "There was greater resistance to the girls having to go to school, because their parents felt that learning to read and write would be of no use to them, since in the daily activities of the family it is the males who deal with problems and handle relations with outsiders."

All these daily activities converge around 5:00 P.M., when the herd is driven back to the corral by the women and the men return from the field. The sky slowly darkens and the wind picks up a bit, spreading a blanket of cold night air across the valley. The chickens strut to their roosts after a respectable amount of cackling and flapping of wings. The call of the pichiko begins as *ch'issin ch'aska* ("evening star") glimmers into view above the ridge of mountains to the west.

Dinner is now prepared (often jointly by men and women) and eaten. The evening meal is similar to breakfast, the major difference being the infrequent appearance of soup. There follows a period of visiting and drinking chicha or *trago* (cane liquor), which lasts until 8:00 or 9:00 P.M., at which time most people go to bed. The single doorway of the dwelling is now closed and will remain closed until morning.

In general, it is not considered good to be out at night. Young men occasionally pasture sheep or cattle late at night, or very early in the morning, but this is not very common. The night is populated by malevolent beings, which makes it an unsafe time to be outside of the house (J. Núñez del Prado 1970: 82-89, O. Núñez del Prado 1973:39, and Wagner 1978:71).

The sun is most commonly used to reckon time during the day (chap. 3). At night, the moon is observed when possible (chap. 4), but the most reliable method for determining time at night is by the rise, culmination, and set of either three or four bright stars or planets (chap. 8). The rising of the bright star *pachapacariq ch'aska* ("dawn of the earth/time star") between 3:30 and 4:00 A.M. is followed shortly by the crisp call of the pichiko.

The Cycle of Weeks

The Spanish chronicles contain some evidence that before the Conquest, the Inca "week" was of eight or ten days (Zuidema 1977a:229), and Rosa Gomarra-Thompson (personal communication, 1976) has found an eight-day agricultural week still in operation in a few communities on the coast of Peru. In most present-day communities, however, the week is counted as the seven-day period of the European calendar, and the names used for the days of the week are also European (Spanish). Beyond the arcadian period, the week is the next longest period of time by which activities are organized and through which they cycle.

From Monday through Friday, the usual pattern of activities resembles that described for the daily cycle. The actual duties vary for the men in accordance with the progression of the agricultural cycle; for the women, it is not so much a variation of duties as a shifting of their location—changing pasturelands. One event interrupting daily activities is the nearest large market, located in Urubamba in the Vilcanota Valley and held every Wednesday. Families often load a burro with produce for exchange at the market (sacks of corn, grain, or potatoes) and set off early in the morning for the two-hour walk down to Urubamba. Occasionally, two or three women travel alone to Urubamba, but men almost always make the trip when produce from the family storehouse is taken to market for barter. This reflects a general pattern whereby men barter and sell *outside* the community and women are responsible for transactions *inside* the community. The principal exception to this pattern in Misminay is that the largest of the two *tiendas* ("stores") is owned and operated by a man. The smaller tienda is run by a woman, although it is said to be the tienda of her husband.

A small market held on Mondays in Maras is attended irregularly and usually by women. In addition, a bus leaves early every morning from Maras to Cuzco and returns in the late afternoon. This bus is commonly used by market women from Maras, but people from Misminay (usually men) occasionally travel to the large Saturday market in Cuzco. Thus, most travel outside the community involves the marketing activities of Monday, Wednesday, and Saturday.

Saturdays are treated like weekdays, with the above marketing exception. During the rainy season, when the footpaths are running with thick brown mud and the fields are soggy, men either work in fields adjacent to the houses or help with pasturing the livestock. This pattern is intensified on Sundays, when almost no work is done in the fields. However, the livestock must still be pastured, and men and women often spend Sunday together pasturing the herd either in the mountains or in Moray.

Sunday is also a day of rather heavy and continuous drinking. Often on Sundays during the rainy season, when movement abroad is difficult, the livestock are kept in the corral and fed weeds and grasses cut from around the house. On such days drinking begins early in the morning and lasts until late afternoon. A few women in the community may prepare large urns (*raki*) of chicha; they invite relatives to drink for free and sell the rest. A two-quart jug of chicha is sold for two or three *soles* (one sole = ca. one cent). Trago can be purchased in either of the two tiendas for about thirty soles a quart. Tienda owners haul the trago up to Misminay from Urubamba in large gasoline cans strapped to the backs of burros. A man who does not own a tienda might try to make extra cash by buying a gasoline drum of trago and selling it at a slightly cheaper rate than the tienda price. Thus, men attempt to increase their capital by bringing trago in from the outside, while women do so by preparing and selling chicha. It would be as inappropriate for a man to make and sell chicha as it would be for a woman to purchase and sell trago. In one respect, this may reflect a difference in whether the substance is *made* within the community and sold (female) or *bought* outside the community and sold (male).

In Misminay, then, weekly activities are vaguely divided into periods of five days and two or of six days and one in relation to the scheduling of male agricultural duties. However, pastoral duties carry on through the weekend, and thus the division of the week is ultimately not a division into days of work and days of rest. Rather, the week is best thought of as a period in which Sunday is designated a nonagricultural day and pastoral duties are shared by men and women.

In addition to differences in the patterns of working and marketing on specific days of the week, the seven days are classified differently with respect to other activities. Mondays and Fridays are said to be bad days for travel or for *any* movement outside the house. This belief was stated to me explicitly several times, but I was never aware of a decrease in movement, not in terms of work in the fields or of travel to the Monday market in Maras, on either of these days. The belief may be reflected in *anxiety* over movement rather than in an actual curtailment of movement, although I was never aware of such anxiety.

Divination by means of coca and maize kernels is common in Misminay. Divining the location of lost or stolen articles, performed mostly with coca rather than maize, does not seem to be restricted to certain days—when something is lost or stolen, an immediate attempt is made to divine its location. However, there are "good days" and "bad days" to consult a diviner-priest about a specific personal concern such as a marital problem or the best time to make a transaction. Tuesdays, Wednesdays, and Fridays are good for divinations of this type; Thursdays and Sundays are bad days (I did not hear a bad/good classification for Mondays or Saturdays). The grandfather of one of my principal informants divines with maize kernels. I asked him on three different occasions to divine for me concerning a personal matter; he refused on two consecutive Thursdays but consented on Wednesday of the third week.

The Duties of the Months

Just as the names and number of days in the week in Misminay are derived from the European calendar, so too are the number of weeks in a month. This means, of course, that months are counted as four weeks or about thirty days, which is approximately the lunar synodic period (29.5 days). The analysis of the pre-Spanish mode of reckoning monthly periods is one of the most difficult problems in the study of the Inca calendar system. The only systematic attempt to deal with this problem is the work of R. T. Zuidema (1977a), whose principal approach has been to study the cycles of religious ritual, agricultural duties, and political activity and to correlate these with various lunar and solar time periods (e.g., the synodic and sidereal lunar periods, solar periods marked by the equinoxes, solstices, and the zenith and nadir sun). Various aspects of Zuidema's calendrical analyses are discussed later.

Certain activities and behavior in Misminay have a relation, either in timing or duration, to the month or the moon or both; this equation of the moon and the month is explicit in Quechua (*quilla* = "moon" and "month"). We are concerned at the moment with monthly behavior; chapter 3 deals with lunar (monthly) astronomy.

A look at the cycle of agricultural activities in Misminay (fig. 1) shows that the sequence of duties is roughly correlated to monthly units of time. Figure 1 is organized to show the relationship between the months of the year and the progression of agricultural duties from planting to harvesting. The calendar starts with June 24, when the festival of San Juan is celebrated. On this date agricultural and climatological predictions are made in Misminay for the coming year. The celebration of the day of San Juan is very probably related to the pre-Spanish celebration of the June solstice (Morote Best 1955 and Orlove 1979:92).

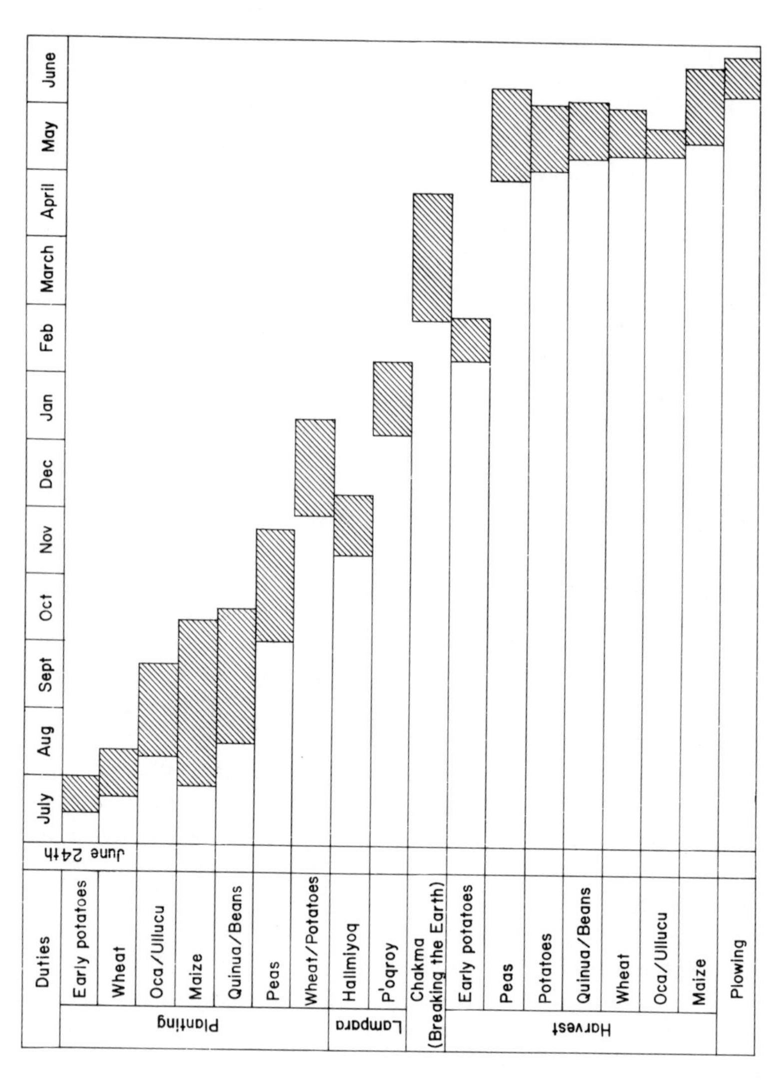

Fig. 1. The Duties of the Months

The figure is not intended to limit the possibilities for representing the calendar in Misminay—calendars could be constructed based on the organization of time and activities in pastoralism, house construction, or weaving. However, the agricultural calendar is probably the most reasonable way to express the progression of monthly activities because the existence of Misminay is so dependent on the success of the crops; all other activities are organized around the agricultural cycle.

Not every step in the sequence of agricultural duties is fixed with respect to a specific date; thus, the association between a particular agricultural duty and a particular month is approximate. Once the decision to begin planting has been made on the basis of divination and general consensus, the cycle is set in motion and one task follows another according to the understood sequence of duties, the different requirements of various crops (such as the need to hoe or to irrigate), and the length of time required for each task. For this reason, individual tasks are not arbitrarily bound to fixed dates; the two sequences—the calendar and the series of tasks—correlate only at certain critical points.

Although most agricultural ceremonialism in Misminay is connected with maize, the agricultural cycle does not begin with the planting of maize. The first crops planted are *mahuay papas* ("early potatoes") and *trigo* (wheat). Early potatoes are planted in July in small plots high in the mountains. Wheat is planted from mid-July through early August in plots midway up the mountains. This initial period of planting is followed in August and September by the planting of two high-altitude tubers: *oca (Oxalis tuberosa)* and *ullucu (Ullucus tuberosus)*. The next stage in the sequence is the planting of maize (*Zea mays*), *quinua* (*Chenopodium quinoa*, a high-protein cereal grain) and *abas* ("beans," *Vicia faba*) in plots somewhat smaller than an acre. The plots are located in three major areas: (a) near each house about an acre of land is almost always planted in abas with an occasional mixing of quinua; (b) slightly larger maize fields, mixed with some quinua and abas, are in the three small valleys running east-west through the community; and (c) larger maize fields, up to two acres in size, are about a fifteen-minute walk across the plain to the east-northeast, in an area known as Pichincoto.

Besides being mix-planted rather than reserved for a single crop, the first two areas differ from the third in that they are the exclusive agricultural domain of Misminay and Mistirakai. The maize fields of Pichincoto are shared by the incorporated community of Mullaca-Misminay, and almost all work in the fields of this area is done by large work parties usually in *ayni* (reciprocal labor) groups (for a discussion of the necessity of reciprocal kin relations for the exploitation of resources, see Brush 1975 and 1977). Peas are planted in October and November, and the planting sequence ends in December and January with the planting of potatoes and wheat.

Because my fieldwork in Misminay began in October 1976, I was not able to observe the maize-planting ceremony of September 26. I was told, however, that maize planting commences with the burial of a *despacho* ("offering"), one component of which is the fetus of a llama. This practice has been recorded elsewhere in the Cuzco area (Dalle 1969:139-140 and 1971:34, Morote Best 1955, and Nachtigall 1975). Maize planting is also accompanied by the drinking of chicha in quantity and, in fact, this pattern of alcohol consumption, both of chicha and trago, occurs at the planting of every crop. (Ralph Bolton 1976, Paul L. Doughty 1967, and Catherine Wagner 1978, have discussed the importance of alcohol and the use of coca in a number of ceremonial and ritual contexts.) In Misminay, I noticed three different patterns in the relationship between alcohol and agricultural work: (a) when planting, chicha is consumed in quantity from the early morning throughout the day; (b) in a number of post-planting tasks such as *lampara* ("hoeing") and *chakma* (breaking up the earth with a foot-plow), chicha is consumed in moderate quantities during work breaks and in the evenings; and (c) during the harvest, no alcohol whatsoever is consumed during the day but limited quantities of chicha and trago may be drunk in the evenings.

The almost complete abstinence from alcohol at harvest time could be related to a practical concern for thoroughness in gathering every bit of the year's crop. This is not to suggest, however, that heavy alcohol consumption during planting is related to an "impractical" concern. Instead, it may be related to general anxiety over the success of the crops, or, on a symbolic level, it may represent the saturation of the body with corn beer (chicha) as a kind of metaphor for the saturation of the earth with seeds: chicha is made by grinding up germinating maize kernels (*wiñapu*); consumed in the form of chicha, these kernels may be thought of as equivalent to the seeds which will germinate in the earth to produce an abundant harvest. This metaphoric equivalence can be extended to the offering, which involves the burial of a "germinating" animal (the fetus of a llama). Catherine Allen (personal communication, 1977; bibliography, see Catherine Wagner) has also suggested that drinking heavily may be a way of "calling up" rain.

One of the major agricultural duties spanning the height of the rainy season, from late November through February, is *lampara* ("hoeing"). Lampara is the process of covering the lower portion of maize and potato plants with dirt pulled up from the furrows with short, crooked hoes (*lampa*). The first period of lampara takes place in late November. This first hoeing, called *hallmiyoq*, is done when the maize plants are about a foot high. The second period of lampara, called *p'oqroy*, consumes most of the month of January. The maize is now about head-high, and the task of hoeing is much more difficult as one must work between the maize stalks instead of over them. The two periods of lampara bracket the planting period for potatoes and wheat (see fig. 1).

For a short period in mid-February, early potatoes are harvested. The balance of the period from late February until the beginning of the harvest in late April is taken up with *chakma*, the process of breaking up the earth with footplows (*chakitaqlla*). This task prepares fields which have lain fallow for two years for planting in potatoes the next year. Chakma is performed by groups of three men working in teams. Two of the men use chakitaqllas to loosen large clods of earth, which are turned over by the third man. The clods of earth are then broken up with pickaxes wielded by young boys and old men.

Chakma is followed by the harvest, which lasts only about a month to a month and a half. The end of the agricultural year comes in mid-June, when the maize, quinua, and abas fields are plowed by teams of two bulls pulling long plows (*arados*) guided by a man. The days of June are crisp and cool, the earth is brown from the lack of rain, and the sky is filled with fine, dry dust rising up behind the teams of bulls.

This, in an abbreviated form, is the series of duties comprising the agricultural cycle from planting to the plowing under of the maize, quinua, and abas stalks. The calendar of activities shows that most agricultural tasks last about a month to a month and a half. Therefore, it is reasonable to assume that the cycles of the moon are used for timing and noting the duration of the sequence of duties. Analysis of lunar astronomical data will show that certain crops (especially tubers) are planted in relation to the phases of the moon, whereas the planting of other crops (especially maize) is related primarily to the sun. Thus, the agricultural calendar correlates *at least* three phenomena, of which two are astronomical (the solar and lunar cycles) and one agricultural (the sequence of tasks).

How these three phenomena are correlated in the calendar is illustrated by considering one of the periods when certain agricultural tasks must be performed at fairly specific times in order to help insure the survival of the crop. Maize, for example, has a growing season of around eight months or, if irrigated, of around six to seven months (Mitchell 1977:47). At the altitude of Misminay, the worst period of winter frosts extends about two and one-half months, from mid-May through July. This means that maize should not be planted before the end of July or after the end of September (see fig. 2). Chapter 3 analyzes a solar observational method for timing this critical two-and-one-half-month planting period in Misminay, but here the question is whether the calendar system now in use in Misminay—the calendar of Catholic saints' days—has in any way been adapted to this critical agricultural period.

Most Andean communities recognize a series of some four saints' days which are celebrated by the community as a whole. The saints' days are part of the community's definition of itself as a distinct group of people (see Brownrigg 1973). For example, one of the days celebrated is usually the day of the traditional founding of the community. In practice, however, there is often a con-

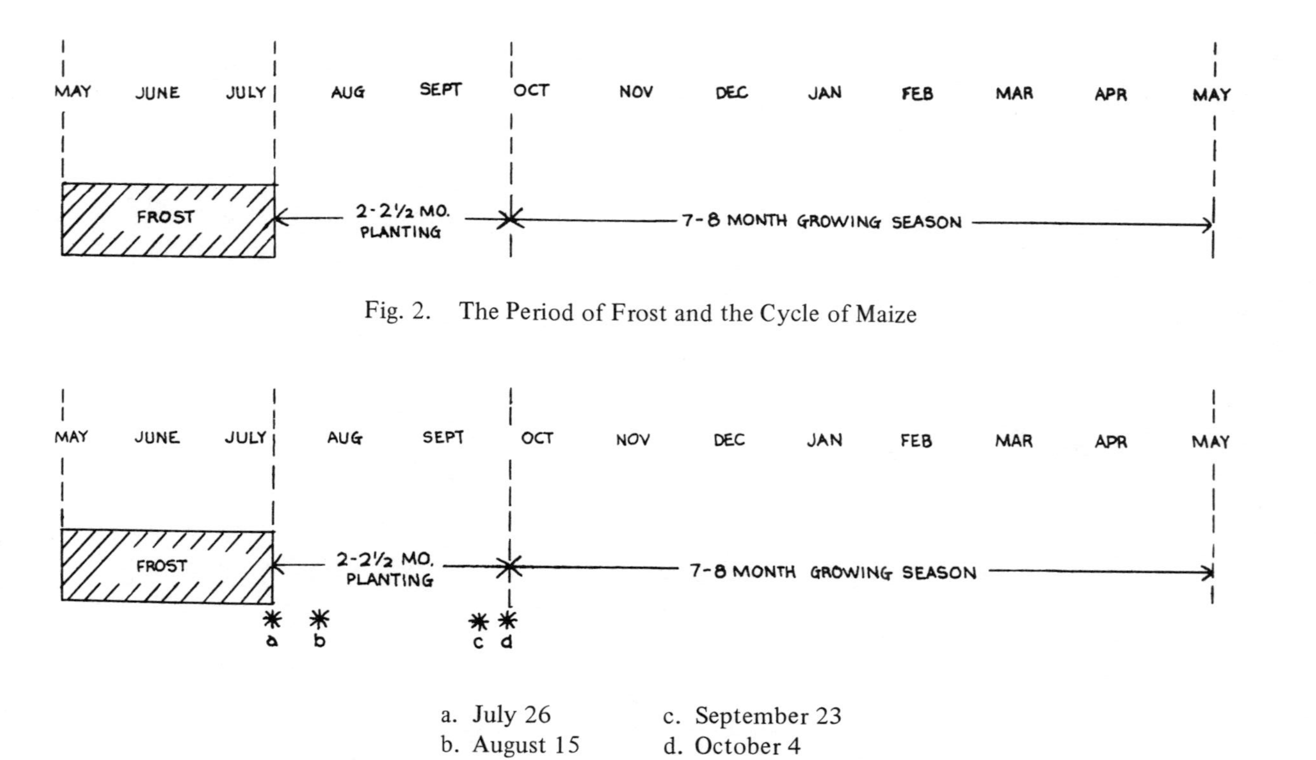

Fig. 2. The Period of Frost and the Cycle of Maize

a. July 26
b. August 15
c. September 23
d. October 4

Fig. 3. Saints' Days and the Planting of Maize

siderable overlapping of saints' days among communities within the same area. In Misminay, the four saints' days are July 26 (Santa Ana), August 15 (Mamacha Asunta), September 23 (Mamacha de las Mercedes), and October 4 (San Francisco). If this sequence of saints' days is applied to the data in figure 2, it very closely brackets the period when maize can be safely planted (see fig. 3).

Seasonal Divisions and Activities

The year can be divided into two parts—the rainy season and the dry season—as well as into the monthly periods related to agricultural tasks. The wet/dry divisions within the cycle of the four seasons is approximated in figure 4. The

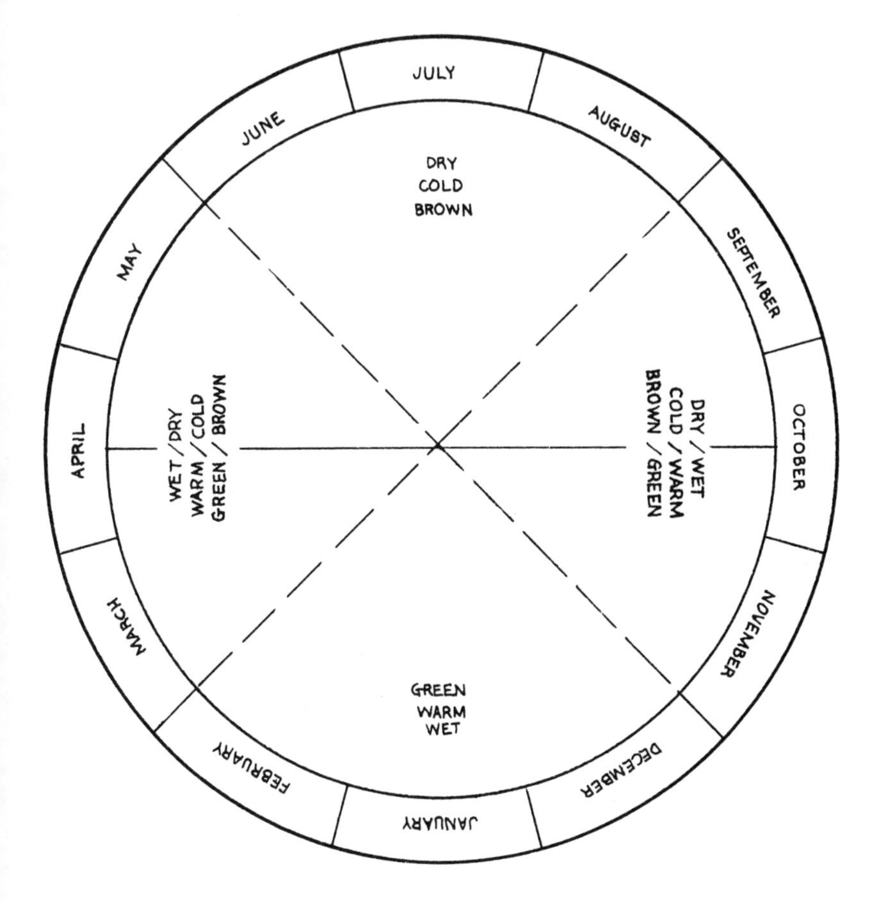

Fig. 4. The Seasonal Cycle

six-month period of the dry season is composed of one three-month period of dryness between two periods of relative dryness that each last one and one-half months; the rainy season is similarly composed of rainy months and relatively rainy months. In Misminay, the differences between cold/dry and warm/wet are believed to be related to differences in the size of the sun (chap. 4).

The two seasons are associated with different patterns of activity, several of which relate to women. During the rainy season, grass for grazing is abundant. However, because much of the land within and around the community is planted in crops, pasturing is done in areas farther away from the community and in the fields which are lying fallow. This pattern of dispersed rainy-season grazing means that the women must drive the livestock either higher up into the mountains or lower down near Moray and in the plain to the east. As the dry season approaches and the harvest begins, fields nearer the village slowly become available for grazing. By the end of the harvest, during the driest months of the year, the livestock (especially cattle) are grazed in the fields adjacent to the houses and in the three valleys running through the community. Not only is the pasturing more convenient, but the cattle are an important source of fertilizer for the maize fields.

Thus, the annual pattern of grazing can be characterized as a rhythmic expansion and contraction centering on the community. Because the seasons are also closely related to the sequence of agricultural duties, such as planting, chakma, and harvesting, the pastoral cycle is integrated with the agricultural cycle. These relationships can be expressed as a correlation of male (agricultural) time and space with female (pastoral) time and space. Lionel Vallée has noted also that the communal and noncommunal work projects are related to different seasons. For example, most communal projects take place during the dry season (Vallée 1972:247-248).

Weaving is another activity of women which seems to depend on the season. In Misminay, and in most communities in the Cuzco area, it is done out of doors on a backstrap loom. I have never seen weaving take place indoors. Because of this apparent outdoor requirement, weaving is normally done during the dry season. In the community of Sonqo, Catherine Allen (personal communication, 1978) was told explicitly that the weaving month is August. This is not to say that weaving cannot be done at other times of the year, only that it is usually performed during the dry season. Although the dry season and the daytime are often classified symbolically as a male period, weaving practices would allow them to be classified as periods of female activity.

The rainy season and dry season are also associated with different types of sounds: drums and *quenas* ("flutes") are related to the rainy season; wooden *pututus* ("trumpets"), to the dry season. These associations are found in the following ways.

During the two hoeing periods of November and January, groups of men often work all day in reciprocal labor (*ayni*) in the maize fields in the plain near Pichincoto. In the late afternoon, between sunset and darkness, the men return in single file to the village. As they leave the field, they begin beating drums and playing flutes. The music continues as the group passes through the community and into the house of the man for whom the lampara was performed that day. The music becomes louder and more excited as large glasses of chicha are distributed among the musician-workers by the woman of the household. The husband then produces a couple of bottles of trago, which is drunk in shots from small clam shells. The music and festivities continue until 9:00 or 10:00 P.M. On one such occasion, I was told that drum and flute music helps the maize grow. (William P. Mitchell [1977:52-53] has discussed the association of drum and flute music with the cleaning of irrigation canals in the community of Quinua; drums and flutes are the only instruments used there during the irrigation festival held at the beginning of the rainy season.)

In many Andean communities, the word *pututu* refers to large conch-shell trumpets which, blown through the dorsal end, produce a deep, resonant sound. Pututus often call the villagers together for a community meeting early in the morning. In Misminay this is done by blowing a small whistle, the metallic screech of which must be infinitely more unpleasant to awaken to than the moan of a conch shell. The pututu of Misminay (see fig. 5) is made of two pieces of light, hard wood which come from the Urubamba Valley. Although wooden, its sound is similar to conch shells but not quite as deep. In Misminay, the pututu is blown in the morning at the time of the wheat harvest in April and May; they are not blown indoors in the evening as are drums and flutes during

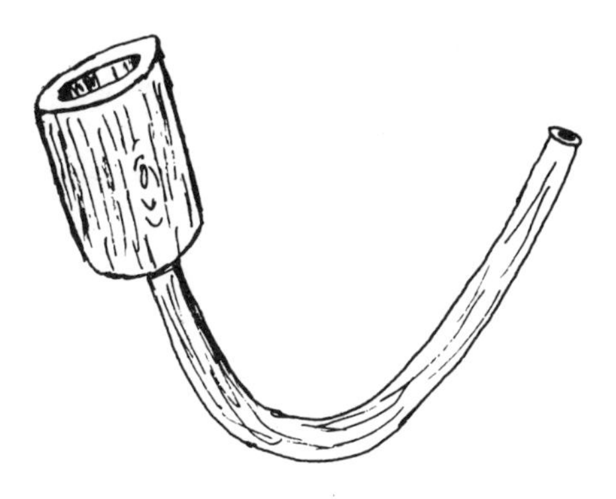

Fig. 5. Wooden *Pututu*

the lampara of maize. In nearby Maras, the task of transporting and storing the harvested wheat is done to the sound of the pututu, the playing of which Patricio Arroyo (1974:5-6) reports is ordered by the *chacrayoj* ("the one in charge of the fields").

Thus, two types of music/sound are related to different crops and seasons: drums and flutes to the lampara of maize at the height of the rainy season (November and January); pututus to the wheat harvest during the early part of the dry season (April and May). It is important to note that the divisions and passage of time are related to different crops and activities, and that these in turn are related to the different senses. Since the astronomical reckoning of time depends primarily on vision, the above material suggests that the total perception of time and space will involve the union of all sensual perceptions of change in the environment.

Laymi de la Comunidad (Crop-Rotation Cycles)

At each stage in our description of the calendrical ordering of activities in Misminay, we have found that shorter cycles of time and sequences of activities are combined to form longer cycles and sequences. The same process occurs with the period of the year and the annual sequence of duties: the two are joined to form multiannual cycles. The easiest way to observe and describe these periods is to use the example of *laymi,* the cycle of crop rotation and fallowing (see the discussions of laymi in Albó 1972:784-786 and Vargas 1936: 225). The two most important crop rotation cycles in Misminay are of wheat, a three-year cycle, and potatoes, a five-year cycle. These multiannual cycles are important for this study, because they account for regular shifts in the orientation of the community with respect to the fields under cultivation; because they produce yearly differences in the specific agricultural tasks performed; and finally, because careful study of the multiannual organization of activities yields important clues concerning the incorporation of long astronomical cycles and periodicities in the calendar system.

The three-year-wheat rotational cycle operates as follows:

Year 1–plant wheat
Year 2–fallow/pasture
Year 3–chakma (break the earth)
Year 4/1–plant wheat

The five-year rotational cycle of the potato fields operates in a similar pattern:

Year 1–plant potatoes
Year 2–plant barley
Year 3–fallow/pasture

Year 4–fallow/pasture
Year 5–chakma

Year 6/1–plant potatoes

It should be pointed out first that these rotational cycles not only provide the fallowing time necessary for the replenishment of minerals exhausted by the cultivation of potatoes (pasturing years equal fertilizing years) but simultaneously integrate the cycling of agricultural lands with that of pastoral lands. This results in a further integration of female and male activity cycles.

In speaking with informants about the rotation of crops, I was given the names of one group of fifteen areas and another of seven areas around the community through which *potatoes* are cycled. The names of the fields, in the order in which they were given, are listed in table 1.

Table 1. Potato-Field Names

Quisqamoko	Wañumarka
K'otin	Quisqamoko
Huaylanka	K'otin
Ch'uñomasana	Huaylanka
Sonqo	Sonqo/Ch'uñomasana
Sacrachayoq	Huarchoq
Anaresq'ana	Conchayoq
Chacracasa	
Llabichayoq	
Yanaurqo	
Conchayoq	
Huasauju	
Karnisayoq	
Ancahuachana	
Kunukayoq	

The lists in table 1 were given by several different informants (which may account for some of the differences in naming). Most informants, however, seemed more concerned that I record the proper number of areas rather than their specific names.

As mentioned, the two lists in table 1 were specifically related to the five-year rotational cycle of potatoes; nothing was said about separate names for the fields used in the rotation of wheat. However, the two groups of fields, one of seven and one of fifteen, are probably the result of integrating the three-year wheat cycle with the five-year potato cycle.

Table 2 outlines the integration of the two cycles. In the table the potato and wheat cycles begin simultaneously, as though they pertained to two different plots, and the two cycles run through their respective periods until they once again coincide with the planting of wheat and potatoes in the same year in the same two plots. The two epicycles (three and five) complete one longer cycle at the end of the fifteenth year (beginning a new cycle in the sixteenth).

Table 2. The Integration of the Potato and Wheat Cycles

Year	Potato Field	Wheat Field
1	potatoes	wheat
2	barley	fallow
3	fallow	chakma
4	fallow	wheat
5	chakma	fallow
6	potatoes	chakma
7	barley	wheat
8	fallow	fallow
9	fallow	chakma
10	chakma	wheat
11	potatoes	fallow
12	barley	chakma
13	fallow	wheat
14	fallow	fallow
15	chakma	chakma
16/1	potatoes	wheat

Thus, it is possible that the presence in Misminay of fifteen areas used for the rotation of potatoes is the result of integrating the three-year cycle of wheat with the five-year cycle of potatoes. This suggests, in turn, that each of the fifteen *areas* represents one *year* in a fifteen-year cycle.

There remains the problem of explaining the seven-area list in table 1. Another look at table 2 shows that the eighth year of the fifteen-year cycle marks the first coincidence of the *fallowing aspect* of the two cycles. Since fallowing periods represent the replacement of agriculture by pasturing (i.e., the replacement of male activities by female activities), the eighth year contains the theoretical possibility of the absence of agriculture. Thus, after the seventh year, the second half of the fifteen-year cycle begins with an orientation toward female activities. In summary, the cycle of crop rotation (laymi) can be used to organize activities over periods of three, five, seven, and fifteen years.

Donald Thompson and John Murra (1966) have described cycles of two and eight years in connection with bridge building. R. T. Zuidema (1977a:230) has discussed a sixteen-year cycle reported in the chronicle of Pedro Sarmiento de Gamboa (chap. 31). Since sixteen solar years are equivalent to ten Venus years ($16 \times 365 = 10 \times 584$), the cycle recorded by Sarmiento provides us with one possible astronomical framework for the correlation of several multiannual cycles of activities; the correlation, that is, of a long solar/planetary cycle with an integrated set of bridge-building, agricultural, and pastoral cycles.

2. The Organization and Structure of Space

One afternoon, I walked from Maras to Misminay with an old man with whom I had become well acquainted. Midway along our walk, we stopped for a few minutes and sat down in a ditch to shelter ourselves from the cold wind. The inevitable bottle of trago was produced, and the wind became warmer. He asked how my work on astronomy was going, and I told him that I still felt completely ignorant. I then asked him if he thought that I would ever understand the sky and the stars. He thought for a minute, and indicating the land around us with a wide sweep of his arm, he asked me if I understood the land and the community yet. When I said that I did not, he drained another cup of trago and asked how, then, could I possibly hope to understand the sky. In keeping with his comments, the following analysis of the astronomy and cosmology of Misminay begins with a description of the topography of the community, not by way of making a methodological statement or simply giving another introduction but because one *cannot* understand the sky without first understanding the earth.

What follows is an analysis of astronomy and cosmology only as they pertain to the community of Misminay. Comparative data from other communities will be discussed, but the "system" of celestial observations and of the integration of the terrestrial and celestial spheres applies to the topographic organization in Misminay.[1] Specific cosmological structures and organizing principles which may have a much wider distribution will be taken up later.

From various accounts given by several different informants, it is clear that the surface of the earth is thought to be curved. One informant insisted that the earth is something like an "orange floating in a bowl of water." At the northern and southern extremities of the earth are two enormous mountains. These mountains, both called Volcán, stand at the boundary of the earth, the sky, and the cosmic sea, which completely encircles the earth. Through the center of the earth, from the southeast to the northwest, flows the Vilcanota (or Urubamba) River. As we will see later, the Vilcanota River is the major artery for the movement of water collected from the smaller tributaries of the earth back to the cosmic sea, from where it is taken up into the sky within the Milky Way and recycled through the universe. The Milky Way is itself thought to be the celestial reflection of the Vilcanota River.

In addition to "this earth" (*kay pacha*) upon which humans live, there is another world directly below; in Misminay, this other world is called *otra nación* ("other nation *or* world"). Informants insisted that otra nación is distinct from *ukhu pacha* ("internal world"), the phrase often used to describe the Quechua "underworld." Otra nación is the place where the dead go, and the entrance to the other world is located west of the village of Misminay.

In otra nación, everything happens just opposite to the way it happens on this earth; our sunrise is their sunset, our day is their night, and our earth is their sky (see fig. 6). The beings who live in otra nación are red and have wings; they are called *condores*.[2] The only animal in otra nación is the burro, which forms the principal food animal of that world. The opposition between this world and otra nación, then, is consistent with regard to the burro; in our world it is a work animal and is not (to my knowledge) eaten, whereas in the other world it is a staple. The only plant food in the other world is the palm tree.

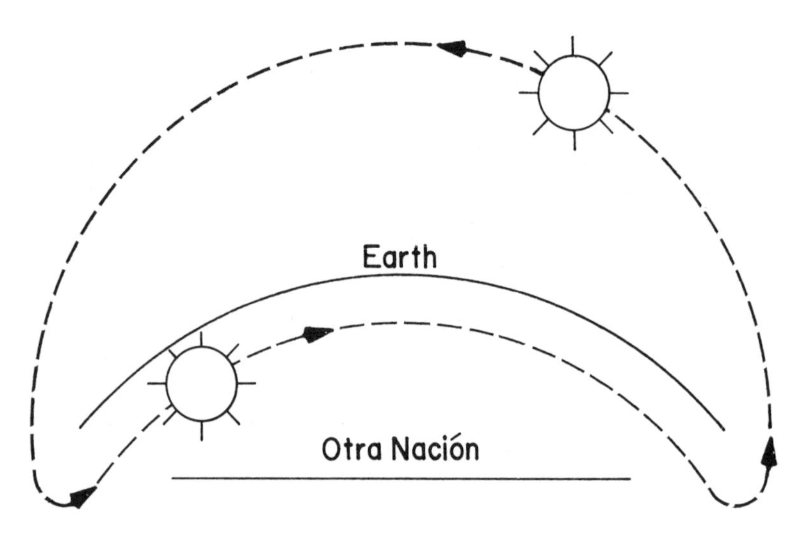

Fig. 6. *Kay Pacha* and *Otra Nación*

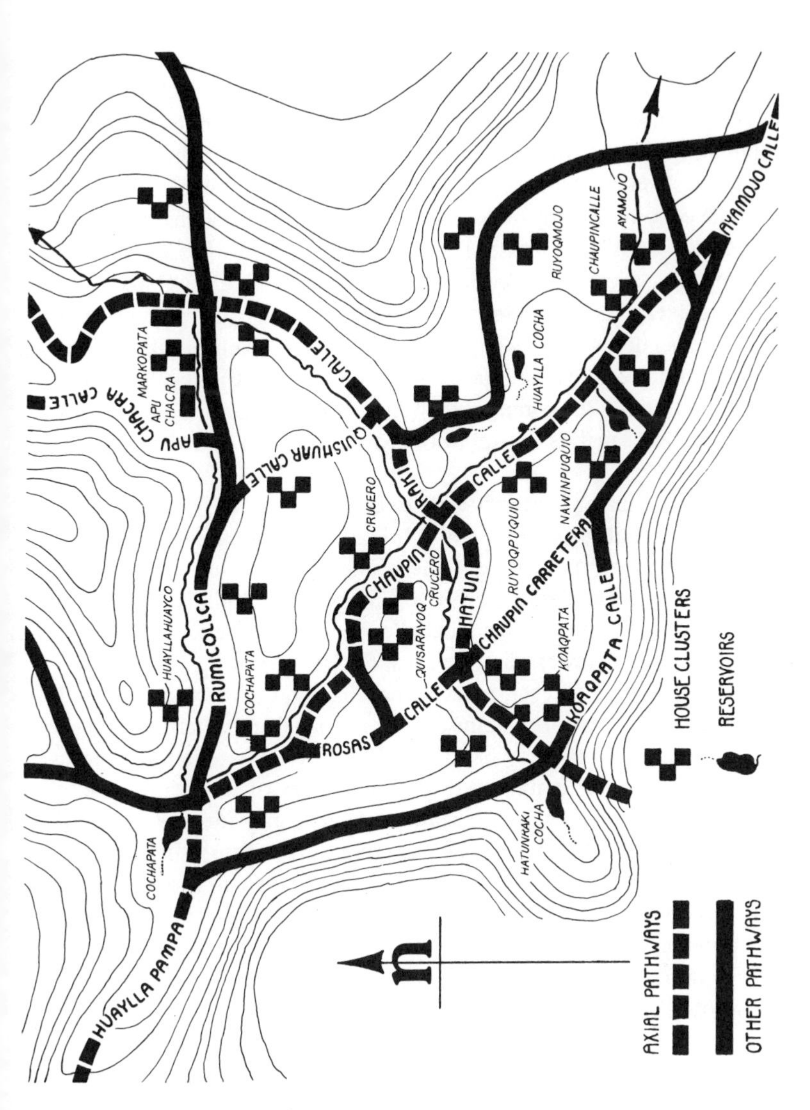

Map 2. Misminay

The Organization and Structure of Terrestrial Space

The community of Misminay (map 2) is made up of some 150 houses, roughly half of which are inhabited (the remaining structures are used as storehouses). The houses are scattered in clusters along three low ridges which run through the valley of Misminay in a generally southwest to northeast direction. The settlement type is midway along a continuum from centralized to dispersed; that is, the community is not clustered around a central plaza as is the case in many highland and coastal communities, nor is it widely dispersed as one often finds in the remote communities of the puna. At the center of Misminay is a chapel called Crucero ("cross"). It houses the three community crosses and is used as the gathering place for religious events and communal activities. Crucero is considered the center of the community not only because it is the principal meeting place, but also because it is the point where the two main footpaths and irrigation canals intersect. Thus, Crucero is the middle point for both water and people moving through the community. The organization of footpaths, reservoirs, and irrigation canals is central to the conception and articulation of space in Misminay.

Footpaths. The major footpath which passes through Misminay is called Chaupin Calle ("middle *or* center path"); it is also less commonly referred to as Calle Chaki ("foot path"). Chaupin Calle is important because it divides Misminay into two parts and because it is the principal pathway linking Misminay to other nearby communities. Chaupin Calle cuts through the center of Misminay in a southeast to northwest direction (see fig. 7).

A ten-minute walk to the southeast along Chaupin Calle brings one to the community of Santa Ana, the administrative, religious, and educational center for the incorporated community of Mullaca-Misminay. Chaupin Calle continues southeastward out of Santa Ana to the community of Mullaca. If one follows the footpath northwest from Misminay, it forks, and one trail goes to the large community of Kaqllarakay (a walk of from thirty to forty-five minutes), while the other leads to the nearby village of Colparay, which is composed of only five or six houses. Thus, Chaupin Calle, the southeast-northwest axis, provides the principal link between Misminay and those surrounding communities which are at more or less the same altitude. Therefore, the "Middle Path" is the axis which socializes Misminay on the horizontal plane.

However, Chaupin Calle is the "center" or "middle" not only because it is the axis for horizontal movement southeast and northwest, but also because it is the axis for the vertical division of space: the half of the community to the south and west of Chaupin Calle is higher than that to the north and east (this is also the effect of Chaupin Calle as it passes through the communities of Santa Ana and Mullaca). Chaupin Calle is therefore the principal axis for the divis-

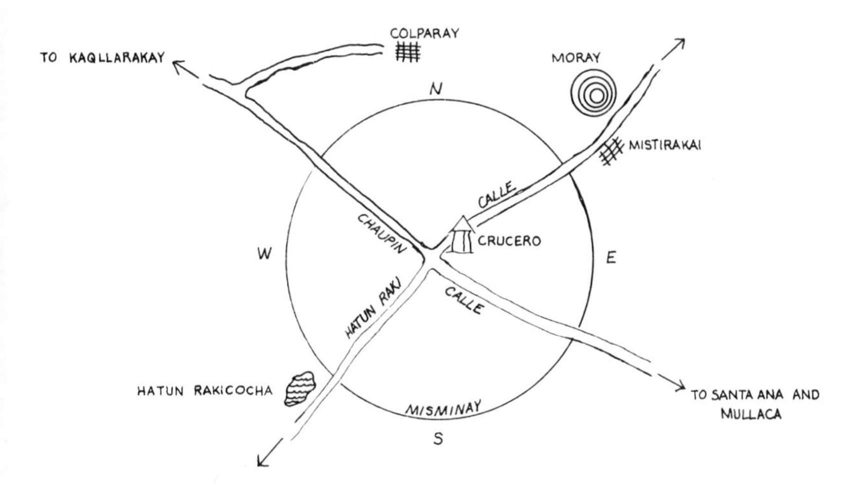

Fig. 7. The Center Path

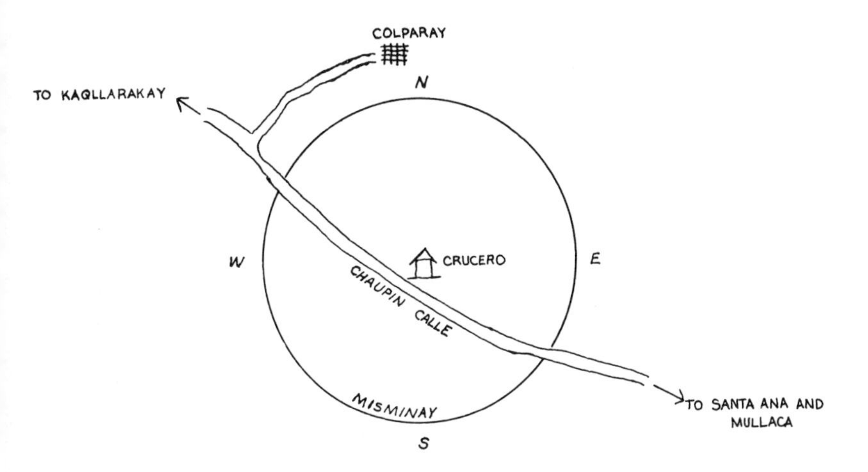

Fig. 8. Footpaths and the Quadripartition of Misminay

ion of Misminay into upper and lower halves (moieties). When someone is walking from Misminay down to Mistirakai, which is located northeast of Crucero, it is common to say that he is going *uray* ("down," a "lower place"). Similarly, someone walking down from Misminay or Mistirakai into the circular ruins of Moray is said to be going *uku* ("interior," or "internal place").

In addition to Chaupin Calle is a footpath that passes through Crucero from the southwest to the northeast (see fig. 8). This pathway is called Hatun Raki Calle ("the path of the great division"). The term *raki* is important for our interpretation of the organization of space within the community. It means (a) "distribution, division"; (b) *tinaja*—a large, open-mouthed earthen jar used to prepare chicha; and (c) "the measure (or division) of land into quarters" (Lira, n.d.). The first two definitions of *raki* will be discussed more fully later. The third defintion, "the measure of land into quarters," is important here. Since Hatun Raki Calle is a single axis which passes through the center of Misminay, there must have been a *prior* two-part division of the community in order to refer to this pathway as the "road of the great division into *quarters*." As noted earlier, the initual dual division is accomplished by the southeast-northwest "Middle Path," Chaupin Calle. In lunar phase terminology, the half-moon is called *chaupin quilla* ("the middle *or* center moon"). Thus, *chaupin* refers to the *division* of objects into two equal parts or to the *union* of two equal units. This again suggests that space in Misminay is first halved by Chaupin Calle and that the halves are subsequently divided by Hatun Raki Calle to produce quarters.

The division of Incaic and contemporary Andean communities into quarters has been well documented (Albó 1972, Fonseca Martel 1976, Palomino 1971, and Zuidema 1964). The two halves or moieties of Inca Cuzco were called *hanan* ("upper") and *hurin* ("lower"), and the quarters were called *suyus* (Zuidema 1964:2-10). In Misminay, I was not given specific moiety names, but because movement from the southwest to the northeast is referred to as going *uray* ("down"), the north and east are likely equivalent to a *hurin* ("lower") moiety, the south and west to an "upper" moiety. In a structural analysis of the relationship between social and geographic systems of classification in the Andes, Jacques Morissette and Luc Racine (1973) have argued that a correlation is often made between natural (e.g., topographic) and social hierarchies. With regard to the upper moiety in Misminay, figure 8 shows that the main reservoir in the southwestern quarter is called Hatun Rakicocha ("the reservoir *or* lake of the great division into quarters"). Thus, the footpath and the reservoir which produce the quadripartite division of Misminay have a common origin in the upper (southwestern) part of the community.

Hatun Raki Calle descends from the mountains, crosses Chaupin Calle at the chapel of Crucero, and continues downward beyond the community of Mistirakai and the ruins of Moray. Thus, whereas movement along Chaupin Calle is horizontal, movement along Hatun Raki Calle follows a diagonal axis through the vertical space of the community. This implies that Chaupin Calle acts as a kind of "fulcrum" for the up and down movement along Hatun Raki Calle. The metaphor of a "balance" or "scale" for the relation between the two axes is not inappropriate since the term *raki* is related to the concept of *áysay*–"to weigh with a balance" (Lira, n.d.).

Reservoirs and Irrigation Canals. In addition to the two major footpaths, which divide Misminay into quarters by intersecting, there is another orientation and division of space, one based on a system of reservoirs and irrigation canals (for discussions of the hydraulic division of space in Andean communities, see Arguedas 1956, Fock n.d., Mitchell 1977, Ossio 1978, Sherbondy 1979, and Zuidema 1978a). Figure 9 is a diagram of the principal reservoirs of Misminay and the *quebradas* ("valleys") into which they flow. In addition to the large canals in Figure 9, a number of small subidiary canals distribute water throughout the fields of the community. The reservoirs are primarily referred to by the term *cocha* ("lake"), but they are also called *pusa* ("conduct, channel, transmit"). The cochas can be said to transmit in the sense that they collect water from streams above Misminay and distribute it throughout the community by way

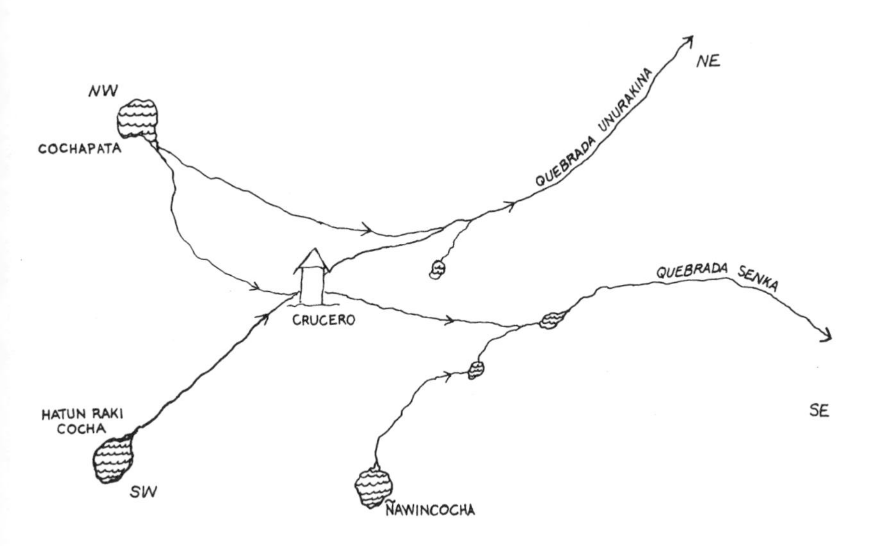

Fig. 9. The Flow of Water in Misminay

of a complex network of subsidiary canals.

The three major reservoirs are located in the northwest (with the chapel of Crucero as the point of orientation), in the southwest, and in the southeast (see fig. 9). The water from these reservoirs flows along two major axes. One hydraulic axis passes from Cochapata in the northwest, through Crucero, and continues downward to the southeast, where it goes underground for about one-half kilometer at a place called Quebrada Senka (*senka*, "to suck liquid up the nose"). The water along this northwest-southeast axis irrigates the fields both within Misminay and along and below Chaupin Calle. This hydraulic axis is also fed by waters from a reservoir in the southeast called Ñawincocha ("eye lake"). The second major hydraulic axis runs from Hatun Raki Cocha in the southwest, through Crucero, and on to the northeast. The water from Hatun Raki Cocha (joined by a small canal from Cochapata) flows in a northeasterly direction into a valley called Quebrada Unurakina ("the valley of the water distributor" or "the valley of the divider of water into four parts"). Therefore, the quadripartite division of Misminay by two intersecting footpaths is complemented by a hydraulic quadripartition (see fig. 10).

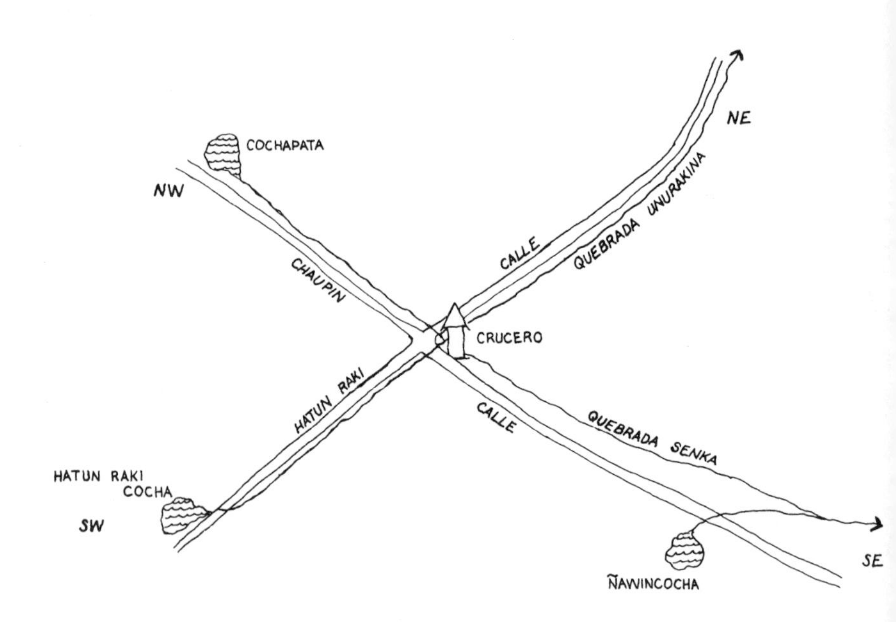

Fig. 10. Schematization of the Footpath and Irrigation-Canal Axes

An interesting comparison for the quadripartite division of Misminay by two hydraulic axes comes from the community of Juncal (Cañar Province) in the highlands of southern Ecuador (Fock, n.d.). Juncal is divided into an upper moiety (Jawa) and a lower moiety (Ura); the intercardinal (northwest-southeast) axis of division between upper and lower Juncal is formed by the branching of a river which originates in the northeast. According to Fock, "it is the partition of the water . . . which gives rise to the socio-political segmentation and it seems to be the distinction between right and left that adds the asymmetrical character to this segmentation" (p. 10). Fock's description of the division of space and its relation to the intercardinal directions is especially interesting: "In Juncal the points of the compass are not abstract points of orientation; they are sectors of a circle such that north means 315°-45° and east 45°-135°; the space from 315°-135° is *Jawa* [upper] and that between 135° and 315° (viewed clockwise) is *Ura* [lower]" (p. 5).

Figure 11, a diagram of the "coordinate" system for the division of space (and hierarchy) in Juncal, reveals a division of space along intercardinal axes which is similar in its structure and orientation to that which occurs in Misminay. In addition, we see the use of the term *ura* (= *uray* in Misminay) for the lower moiety in Juncal. The division of communities into two or four parts along hydraulic axes has also been described for the Department of Ayacucho (Isbell 1978 and Mitchell 1977).

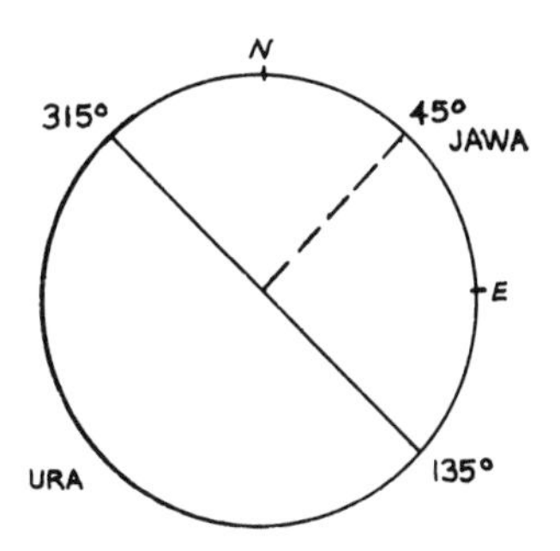

Fig. 11. The Coordinate System in Juncal

The "House Clusters" of the Four Quarters. Misminay was earlier described as a semidispersed community with dwellings grouped into a number of named house clusters.[3] The location of these clusters of houses within the four quarters into which the community is divided is diagramed in figure 12, which is a composite rendering based on information from three different informants in Misminay (see map 2 for more precise locations of these house clusters).

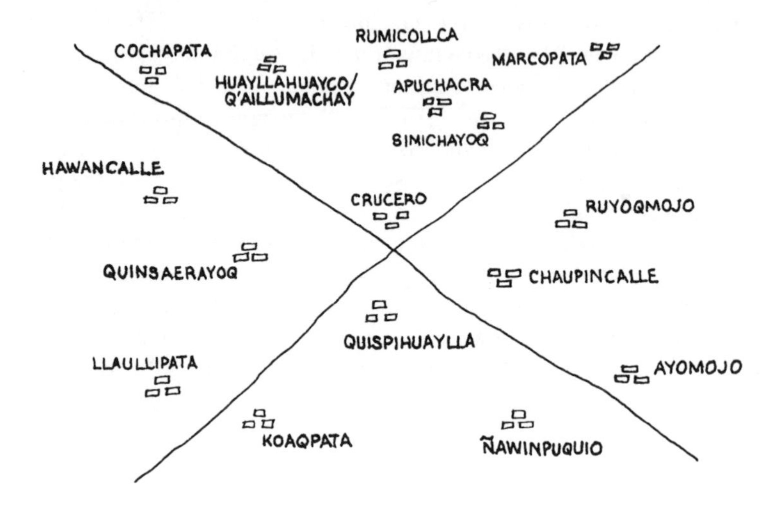

Fig. 12. House Clusters in Misminay

Most obvious is the disproportion in the number of house clusters *per* quarter: north, 7; east, 3; south, 3; west, 3 (total = 16). The largest concentration of house clusters, then, is in the northern quarter, and the north also contains the houses which are related, by name, to the central chapel (Crucero) and to the reservoir Cochapata. In addition, a large field referred to as Apuchacra ("the field of the sacred mountain") is in the northern quarter of the community. Apuchacra is a broad pampa upon which a school was being constructed from 1975 to 1977, and so was the site of a number of communal work projects (*faenas*) during the period of my fieldwork.

The house cluster related to Chaupin Calle is located to the east of Crucero. This would suggest that the Chaupin Calle axis, which divides the community in half from the southeast to the northwest, is a division made primarily from the point of view of the eastern and northern quarters (i.e., from the point of view of the lower moiety).

From these observations, we may conclude that, although the southern and western quarters together form the "highest" moiety of the community in a topographic sense, the north and east form the highest moiety from a "symbolic" viewpoint; that is, within these quarters are the central and axial house clusters–Chaupincalle, Crucero, Apuchacra, and Cochapata. The hierarchical classification of the house clusters is the reverse of the topographic hierarchy (see fig. 13).

A similar kind of reversal of hierarchical classifications has been described by

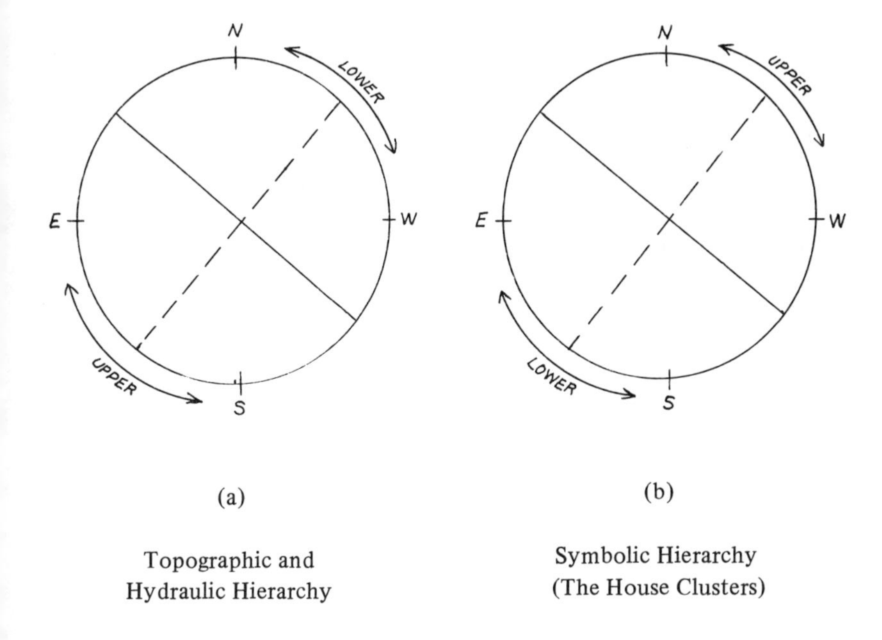

(a)

Topographic and
Hydraulic Hierarchy

(b)

Symbolic Hierarchy
(The House Clusters)

Fig. 13. Two Views of Hierarchy

Christian Barrette (1972) in the central Andean community of Huancaraylla (Province of Víctor Fajardo, Department of Ayacucho). In Huancaraylla, the upper quarter is called Tahuantinsuyu ("the four quarters" or "the fourth quarter"). Tahuantinsuyu is considered to be hierarchically superior to the three lower quarters because it is the quarter of the origin and distribution of the irrigation waters. However, from another point of view, Tahuantinsuyu is classified as the "fourth" quarter with respect to Hurin suyu ("the lower quarter"). Hurin suyu is called the "first" because it is the lowest part of the community; it is the quarter where the earliest inhabitants lived. Thus, Tahuantinsuyu is the highest quarter with respect to the origin of water, Hurin suyu is the highest quarter with respect to the origin of people (for comparative descriptions of the phenomenon of mediation in ecological and symbolic classifications, see Fonseca Martel 1976 and Isbell 1978:145-151).

A similar rationale exists for the reversal of hierarchy in Misminay. One man described the founding of Misminay as follows:

In the beginning, everything was dark and there were neither villages nor farms. Then, a herder came to the area of Misminay from the direction of Maras and began pasturing llamas, alpacas, and sheep on the mountainsides. Another herder came later, and the two men herded near each other. The first man was named Corimaylla, the second Misa Tupac Amaru. The land on Apu Wañumarka proved to be very good both for farming and herding, and the two men stayed there and [presumably with their wives] populated the place.

Therefore, the first inhabitants of Misminay are believed to have come from the northeast, from Maras. (The Incas, whom the *runa* ("people") of Misminay also consider as their ancestors, are said to be now living below ground, inside the ruins of Moray.) In Misminay, then, water originates in the southwest and people originate in the northeast.

A way of further analyzing the hierarchical relationships among the four quarters, and the next step in analyzing the structure and organization of terrestrial space, is to turn to the territory immediately outside the community, especially the nearby mountains which form the visible boundaries of the northern, western, and southern quarters. The horizon lines of these three quarters are in sharp contrast to the broad plain and distant mountains which make up the horizon to the east.

The Visible Boundaries of the Four Quarters. When one stands in the center of Misminay, at Crucero, and takes a 360-degree view of the horizon, the most striking impression is that the horizons to the north, west, and south are very near and high whereas the eastern horizon is low, distant, and spectacular. Figure 14 indicates the magnetic azimuth readings taken from Crucero to those points on the horizon for which I obtained identifications (mountains, distant communities, and so forth).

The recording and presentation of azimuth readings is an artificial, "nonindigenous" method of description, but it will serve the present need of visualizing the relative spacing and orientation of those points on the horizon which are most consistently named and which will be important in order to describe and analyze the conception of space in Misminay from this point forward. An additional aid to visualizing the shape of terrestrial space as viewed from Crucero is figure 15, a composite horizon profile as one would see it from the central chapel. The horizon elevations in figure 15 are within about 3° of accuracy.

The most interesting quarter with which to begin analysis of the symbolism of horizon place names is in the north, where we find the principal *apu* (the name used for the sacred local mountains, the homes of the ancestors) of Misminay, Apu Wañumarka. *Wañumarka* means "the storehouse of the dead": *marka* refers to a storeroom located on the second floor of a house, below the roof; the "dead" referred to are the ancestors of Misminay. The following data reflect the association of the northern quarter with the local ancestors. First,

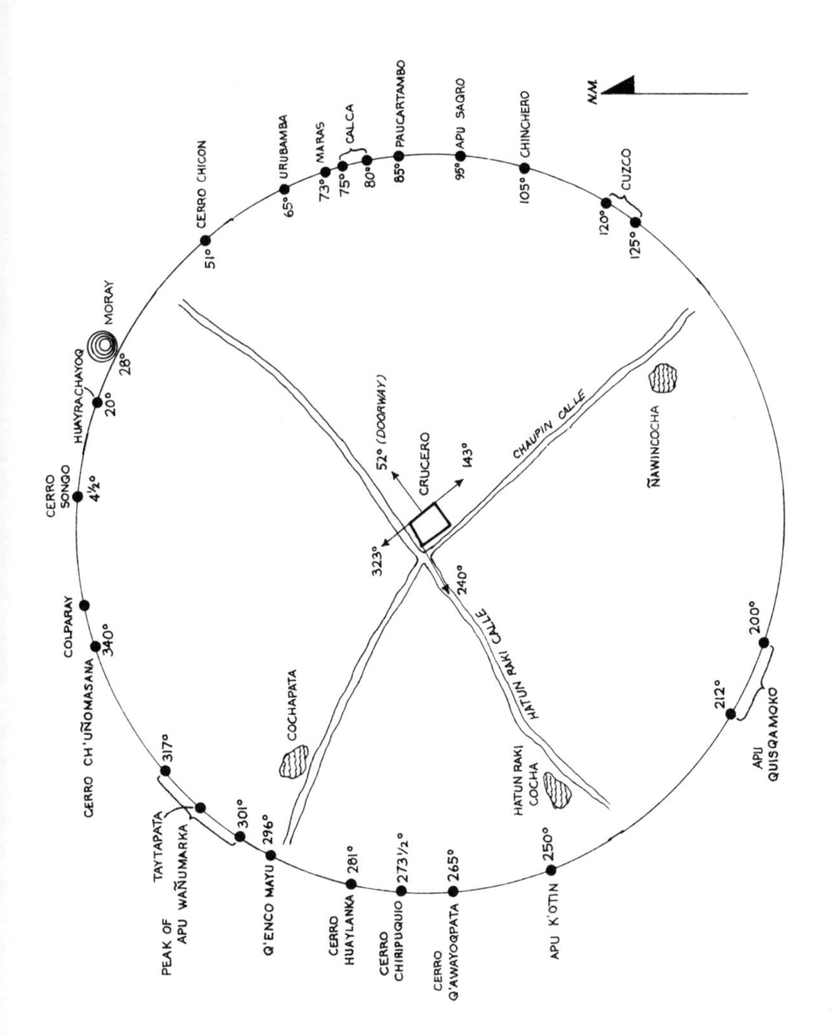

Fig. 14. Azimuth Readings from Crucero

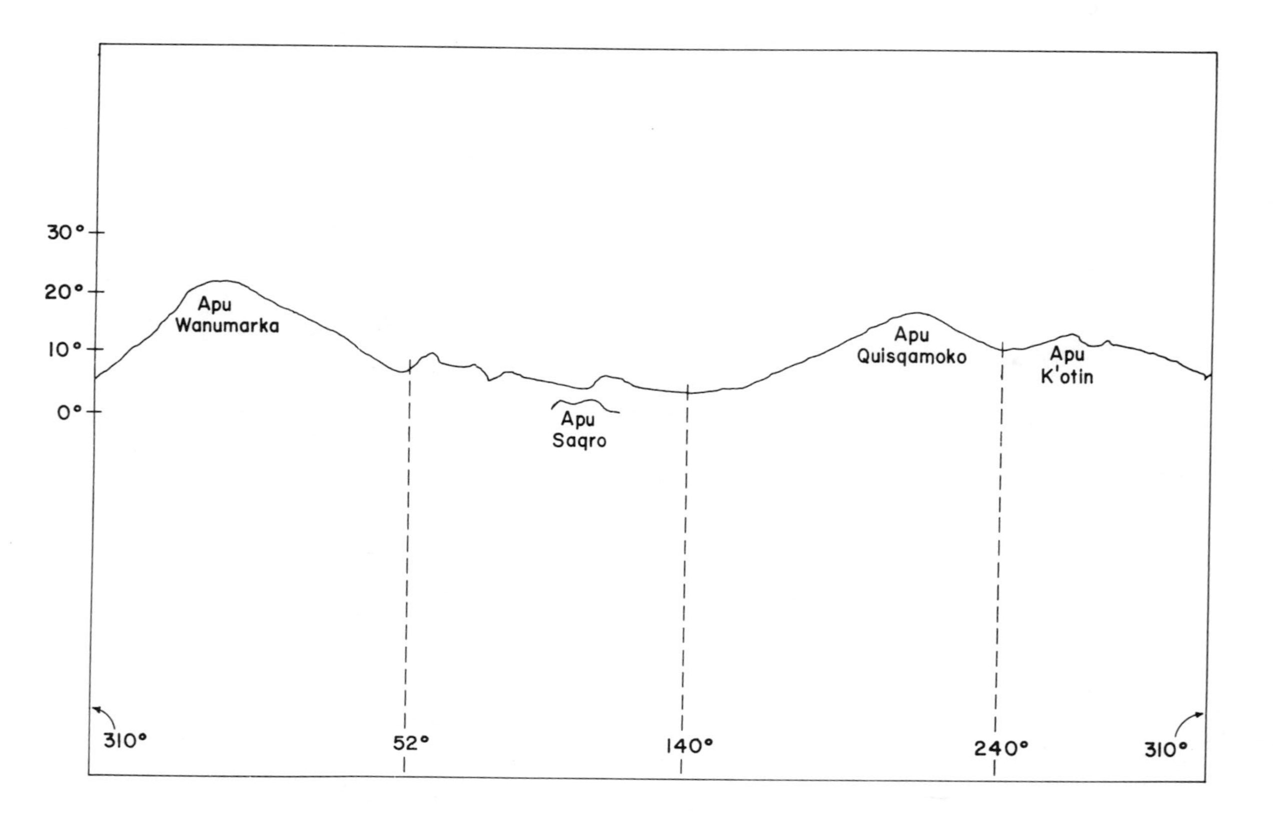

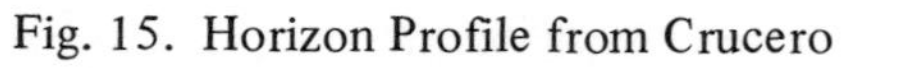

Fig. 15. Horizon Profile from Crucero

the small village of Colparay, located on the northern slope of Apu Wañumarka, is said to be the village where the ancestors of Misminay first settled. Second, at the summit of Apu Wañumarka is a small plain called Taytapata ("the plain of the father" or "the plain of Christ"). I was told that the principal cross of the community of Misminay once stood on this plain ("but who needs a cross," said the informant, "Look! Wañumarka already reaches up to *taytacha* ['Jesus Christ,' 'Lord']!"). A third indication is that, seen to the north from Crucero, there is a ridge where Apu Wañumarka slopes gently to the east, and this ridge is called Ch'uñomasana ("*ch'uño*-in-law"). *Ch'uño* is the name given to potatoes which are alternately frozen and thawed to prepare them for storage, and in Quechua symbolism ch'uño seems to be related to the mummified ancestors (Catherine Allen, personal communication, 1977; see also Oscar Núñez del Prado 1973:37). A fourth consideration is that the Incaic ruins of Moray are located in the northeast. Moray, which is composed of four deep holes shaped like upside-down pyramids, is called upon as one of the community apus (it is, in fact, more like a group of "inverted apus"), and one informant said that the Inca Huayna Capac, the eleventh Inca king, is presently living underground in the largest bowl of Moray.

These data strongly support the notion that the quarter of the north is related to the ancestors and to the concept of origin. It must be emphasized again that the "northern" quarter actually includes the horizon stretching from the northwest, from a point near Q'enco Mayu, to the northeast, near Moray. A later chapter discusses the importance of these two points in the solar cosmology of Misminay; Q'enco Mayu ("labyrinth river") is the setting point of the northern road of the sun, and Huayna Capac, who lives underground in Moray, is one of the names used to refer to the sun.

The "eastern" quarter (from the northeast to the southeast) is significant because the majority of the place names in this direction are related to distant, and, in most cases, subhorizon cities. The subhorizon cities in figure 14 include Cuzco, Paucartambo, Calca, and Urubamba. In the eastern quarter one of the principal visible sites not related to a city is Apu Saqro ("the sacred mountain of the devil" or "the sacred mountain of the cat"). I learned of the existence of this apu only a day before ending my fieldwork. My informant, a young man, said that Apu Saqro is female; she is the wife of Apu Wañumarka and their children are Apu K'otin (in the southwest) and Apu Quisqamoko (in the south). When I asked my informant why I had never heard Apu Saqro's name invoked when making a coca *quintu* (an offering made by blowing across an arrangement of three coca leaves and saying the name of the apu), he said that because she is female, she knows nothing about coca (as noted earlier, Misminay women—ideally—do not chew coca because it is too "hot"; this prohibition, however, is followed only in the most formal of circumstances or when a woman does not care for coca). Since quintus are frequently blown to Apu Quisqamoko and Apu K'otin, I assume that they must be the *male* children of Wañumarka and Saqro.

Apu Saqro is a low, round mountain with a number of agricultural terraces which are of a masonry construction similar to the terraces of Moray. In fact, I was once told that Apu Saqro was the original location of Moray (i.e., that it was called Moray in ancient times). Thus, an ancestral apu in the eastern quarter is equated with one in the northern quarter.

There are a number of interesting place names in the southern and western quarters, but most of the data concerning these sites are more easily discussed in relation to the rising or setting points of celestial phenomena. Discussion of these two quarters is reserved for the section on the organization of celestial space.

Spatial Principles. The first point to note is that the principal divisions of space in the Misminay environment are defined in accordance with topographic features (e.g., the course of the flow of water) rather than in relation to a system of directions based on an external point of orientation (e.g., the pole star as used in the northern hemisphere). Named units of space (sacred mountains, house clusters, and reservoirs) are defined in relation to the two footpath/irrigation-canal axes which intersect at a point referred to as the center, the "cross" (Crucero). Thus, the "coordinate system" for terrestrial space involves a fixed center and two more or less perpendicular axes forming a crossroads. Figure 16 is an approximation of the shape and orientation of this coordinate system. In

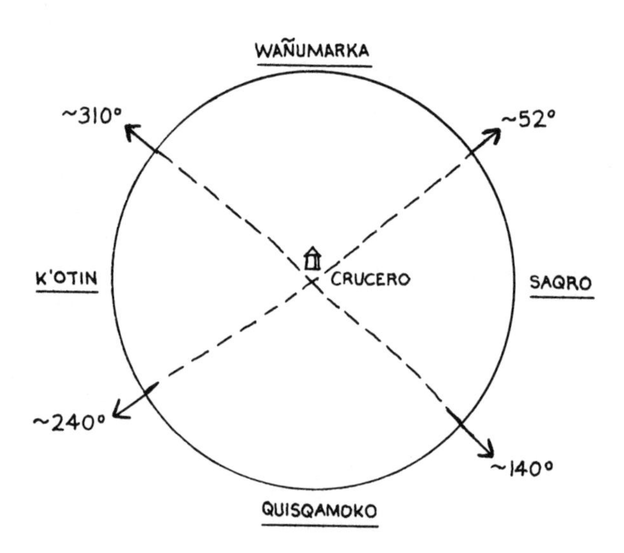

Fig. 16. The Framework of Terrestrial Coordinates in Misminay

view of the fact that the quarters are defined by topographic features (in fig. 16 the quarters are labeled according to their association with a particular apu), we can equate the direction of the quarters with the principal sacred mountains located in each of the quarters (table 3).

Table 3. The Sacred Mountains of the Four Quarters

Directional Quarter	=	Apu Quarter
North (NW-NE)	=	Apu Wañumarka
East (NE-SE)	=	Apu Saqro
South (SE-SW)	=	Apu Quisqamoko
West (SW-NW)	=	Apu K'otin

Secondly, the association of reservoirs and irrigation canals with the division of space inside the community is related to the fact that the three principal reservoirs are located along the footpaths at points which mark the movement from outside to inside the community. Once inside, the footpaths are joined by irrigation canals (see fig. 10). Thus, the footpaths are associated with water (wet) when inside Misminay, but they are dry when outside. This suggests that reservoirs and irrigation canals—and probably water in general—are crucial to the definition of social versus nonsocial space.

The absence of a reservoir in the northeast, at the lower end of Hatun Raki Calle, sets this endpoint, or entrance point, apart from the other three. In this regard, it is interesting to note that the quarter of Apu Wañumarka, in the north, is unique among the quarters because it contains a water *source* at one extreme (Cochapata in the northwest) and a water *receiver* at the other (Moray and Quebrada Unurakina to the northeast).

Topographic and symbolic relationships found among the four quarters are listed in table 4. These topographic and symbolic associations suggest an oppo-

Table 4. The Topography and Symbolism of the Four Quarters

Apu Wañumarka (N)	Apu K'otin (W)	Apu Quisqamoko (S)	Apu Saqro (E)
high horizon	high horizon	high horizon	low horizon
near horizon	near horizon	near horizon	distant horizon
water source/receiver	water source	water source	water receiver
male	male (?)	male (?)	female
father (& ancestor)	child	child	mother
lower	upper	upper	lower

sition between Apu Saqro and a triad composed of Apu Wañumarka, Apu K'otin, and Apu Quisqamoko. This, as we have seen in the sexual and kinship classification of the four apus, is an opposition of a father with his children to their wife/mother. A similar kind of 3:1 disunion has been described in the structure of Andean kinship and the formation of kin-based groups (Zuidema 1977b:247). In addition to this 3:1 classification of the apus, we also find a dual division between two upper apus and two lower apus.

The Organization and Structure of Celestial Space

With the structure and organization of terrestrial space established, we can now turn to celestial space and analyze what relation, if any, exists between the structure of the earth and the sky in Quechua communities. During Inca times there seems to have been a relationship between cosmological orientation and the orientation of cities in the Andes (Urton 1978b). A similar kind of relationship has been found in the town planning of many other ancient and primitive societies (see Tuan 1974 and Wheatley 1971).

When stepping out of a hut at night, a Misminay runa sees very little of the village (the blessed lack of street lamps). What is visible is the dark, jagged circle of the horizon, the outline of the named apus against the slightly-less-dark night sky. Thus, the ring of the horizon, with its named landmarks, provides points of orientation *from which* lines of sight may be extended into the night sky. The peaks of the apus can be likened to fingers pointing upward, providing the initial information for how one goes about ordering (i.e., looking at) the celestial sphere. However, our imaginary nocturnal runa soon encounters a problem because, although familiar landmarks may provide a starting point for a line of sight extended upward, the line goes only a short distance before becoming lost in a jumble of stars. Thus, the runa must find something in the sky which can be used to extend the line further and still retain its initial orientation. It must be remembered that the Misminay runa, living in the southern hemisphere, has no Polaris, no fixed stellar point from which to extend imaginary lines along the celestial sphere and down to the earth, to provide terrestrial orientations as in our own system of longitude and latitude coordinate lines. Because every point in the sky is in motion in the southern hemisphere, the runa can only extend lines upward from fixed points on the earth.

We have already seen how these fixed terrestrial points along the horizon, the apus, are defined with respect to the structure and organization of community space. With the idea in mind that *the apus are transitional between terrestrial and celestial space*, the sky in the southern hemisphere can be examined to see how lines of sight are oriented, and how they retain their orientation, once they leave the earth.

In virtually every discussion I had with the people of Misminay (or any other community) about astronomy, the Milky Way (= Mayu, "river") figured prominently in defining the position of constellations and in relating two or more constellations in the sky. This same orientational phenomenon was encountered by Christopher Wallis (personal communication, 1976), who carried out fieldwork in the Province of Cailloma (Department of Arequipa).

In an interview with one informant in Misminay, the movement of the Milky Way was likened to the movement of a clock:

Por ejemplo ponemos desde la tarde; ya las seis está marcando, las siete, las ocho, las nueve, las diez, entonces ese Calvario de lo que decimos llama, yutu, esa Mayu. Está andando así poco a poco, poco a poco está moviendo, ¿no es cierto?

For example, beginning from the late afternoon, it marks six, seven, eight, nine, ten; that Calvario [cross of Calvary], that which we call llama, tinamou, that River. It all moves like that, little by little; little by little it moves; agreed?

Later, in discussing with the same informant the position of the four constellations identified as crosses, I asked if they are located in the Mayu or not:

Sí, todo el Mayu está junto no más. Los cuatro suyus no más; una, dos cruces no más está separado. Entonces, dos cruces están juntas con Mayu no más. Uno detrás, uno de encima de su adelante.

Yes, the River is just together; the four quarters; one or two crosses are separated from each other [in the Mayu]. Thus, only two crosses are together with (near) the River. One behind it, one in front (ahead).

The one consistent orientational and structural association made by this and other informants concerns the Milky Way (Mayu) and the division of the sky into four parts (suyu). As phrased by another informant in Misminay:

I. Son cuatro suyus; cuatro suyus hay.

U. ¿En el cielo hay cuatro suyus?

I. Sí. Por ejemplo, en este junio va salir de este, entonces para oeste va salir. Entonces a los cuatro suyus tenemos nuestro Dios que está marcando con las estrellas.

I. There are four quarters; four quarters.

U. There are four quarters in the sky?

I. Yes. For example, this June it [the River] rises in the east, thus it will rise in the east and go to the west. Then, we have the four quarters which our Lord marks with the stars.

Knowing that terrestrial space is also divided into four quarters (suyus), we must now attempt to understand the relationship between the structure and organization of celestial and terrestrial space by answering the following questions: (a) How does the Milky Way, the celestial River, divide the night sky into four parts? and (b) What is the relationship between the quarters of the earth and the quarters of the sky?

The Mayu and the Celestial Quadripartition. The Quechua name for the Milky Way, Mayu ("River"), is easy to understand as a reference to a narrow stream of stars flowing through the dark background of the night sky. In Misminay, however, the designation of the Milky Way as a celestial river goes beyond the mere metaphorical equation of a linear stream of water on the earth with a stream of stars in the sky. One informant in Misminay explained the relationship between the two rivers by etching a line on the floor of his hut; the line, he said, represented the Vilcanota River which is like a mirror reflecting the Mayu, the River in the sky. Therefore, the Milky Way is equated *directly* with the Vilcanota River, which flows from the southeast to the northwest through the Department of Cuzco. But this explanation does not capture all of how rivers are conceived of in the Andes nor of what the analogy between a celestial and terrestrial circulating or flowing stream might entail.

Figure 17 is a top view of a typical Andean river and *acequia* (irrigation canal) system as it flows from right to left. The same river and irrigation canals are diagramed from a side view in figure 18. What the two figures illustrate is that

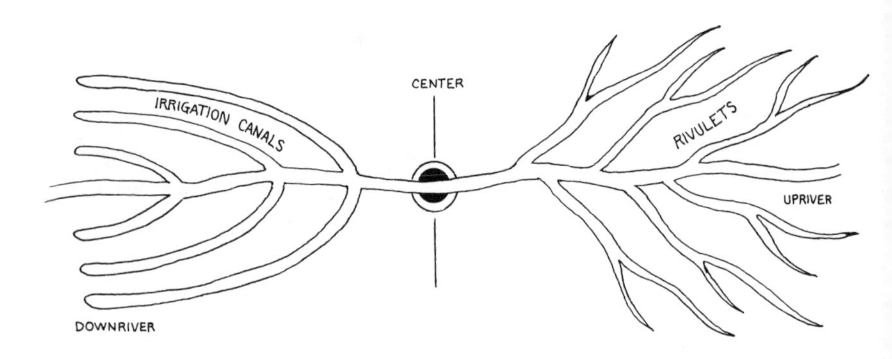

Fig. 17. Top View of River/*Acequia* System

the pattern of tributaries converging with the main course of the river is duplicated, or reversed, after the river is channelized for irrigation purposes. The main course of the river flows from right to left, but on the right, tributaries flow downward and inward to converge with the river, whereas on the left, irrigation canals conduct the water away from the river at a reduced angle. The critical point in this hydraulic system is the point at which the inward motion of branching reverses to outward; that is, at the point where the river attains its maximum input and begins discharging water into the canals. From the point of view of a community living alongside of a river and utilizing it, the river is oriented from this central point which divides the upper river (=inward motion) from the lower river (=outward motion).

Now, how can this Andean conception of a river system be applied to the Milky Way? The Milky Way is a linear stream of stars which divides the celestial sphere into roughly equal hemispheres. Thus, the Milky Way can be thought of as a plane rotating round the earth. However, the plane of rotation of the Milky Way is inclined between 26° and 30° with respect to the north-south plane of the earth's rotational axis (fig. 19).

This difference in orientation between the planes of rotation of the earth and the Milky Way results in a kind of "tumbling" motion in the rise and set pattern of the Milky Way. As the southeastern quadrant of the Milky Way rises, the northwestern quadrant sets; as the northeastern rises, the southwestern sets (this

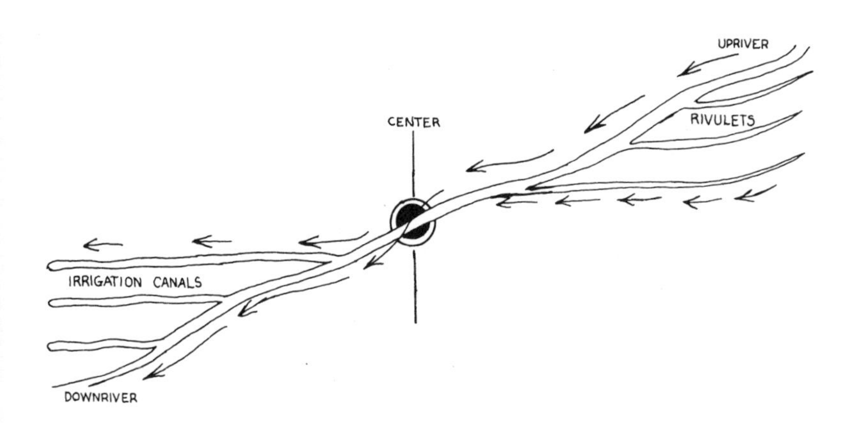

Fig. 18. Side View of River/*Acequia* System

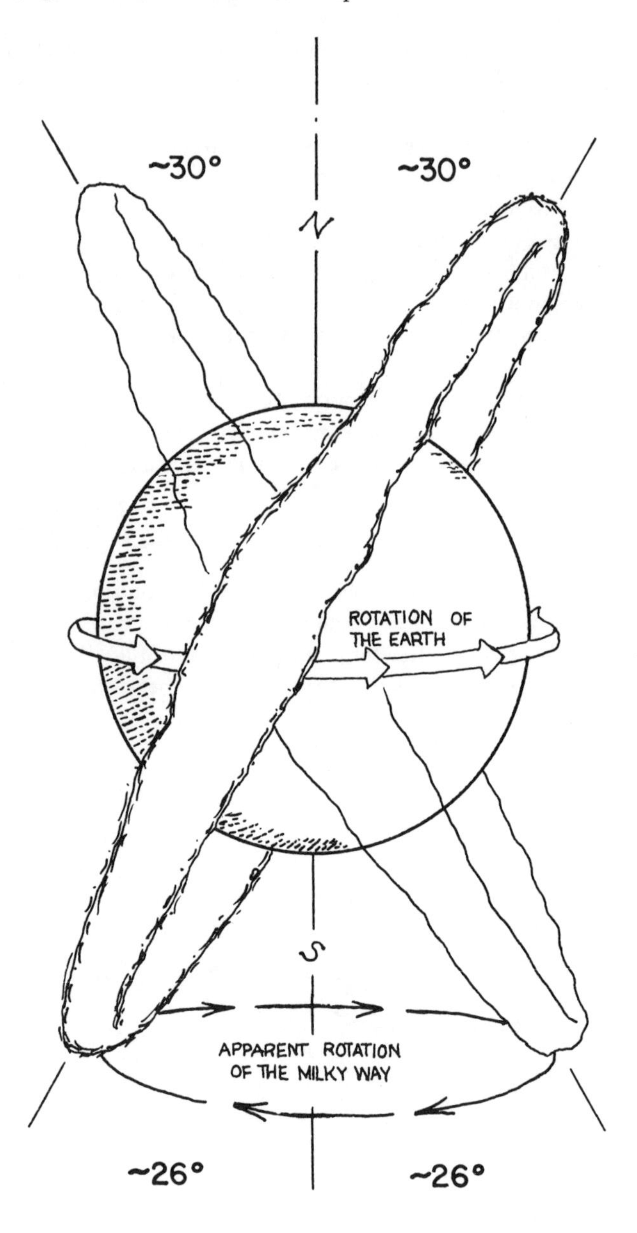

Fig. 19. The Alternating Axes of the Milky Way from a Fixed Point on Earth

motion will be discussed more fully in chap. 8). In Sonqo, an informant described the movement of the Milky Way not by making a smooth circle in the air but rather by tumbling his hands over each other.

For an observer in the southern hemisphere, the point at which the Milky Way comes nearest to the south pole (about 26° from the pole) is at the center of the Coalsack and near α Crucis of the Southern Cross. This point in the Milky Way revolves continuously around the "emptiness" of the unmarked pole. An informant in Misminay, while describing the area around the south celestial pole, commented that this region of the sky *es silencio* ("is silent," unmarked). Since, as we have seen, celestial orientations are made primarily with respect to the line of the Milky Way, the "center" of the celestial River may be taken as the point which falls nearest to—and revolves around—the unmarked south celestial pole; that is, the area near α Crucis and the Coalsack. This area may be referred to as the center in the sense that it separates the hemispheres of the River with respect to the pole, thus separating the eastern rising half from the western setting half.

It is significant to find that the Southern Cross is also used as a point of celestial orientations in other South American Indian cosmological systems (see Wilbert 1975). William Lipkind gives the following description of the cosmography of the Carajá of central Brazil: "Only the southeastern quadrant of the sky interests the Carajá. *Starting at the Southern Cross* virtually every star and constellation visible in the east is a named supernatural. The Milky Way is the shaman's road (1940:249; my emphasis)."

An informant in Misminay described the region of the Southern Cross as the center of the Mayu not because of its revolution around the pole but because it is the point at which *two* celestial Rivers collide. The Milky Way, he said, is actually made up of two rivers, not one. The two Mayus originate at a common point in the north, flow in opposite directions from north to south, and collide head-on in the southern Milky Way. The bright stellar clouds in this part of the Milky Way represent the "foam" (*posuqu*) resulting from the celestial collision. These data indicate that the celestial River has a second center, a "center of origin," in the north.

At the latitude of Misminay (–13°30′), both "ends" of the Milky Way are subterranean for a period of time in their revolution around their respective pole. But since the plane of the Milky Way is inclined with respect to the plane of the earth's rotational axis, one "center" of the Mayu will be above ground while the opposite center is below ground. It must be reiterated here that in Misminay, the earth is thought to be something like "an orange floating in a bowl of water"; it is surrounded by the cosmic ocean (*mar*). We may conclude from these observations that the water in the celestial River neters the celestial sphere when the *northern* end of the Mayu is underground (i.e., when it is in the cosmic ocean) since the *point of origin* of the celestial River is in the north, whereas the *point of union* of the rivers is in the south. This concept is diagramed in figure 20.

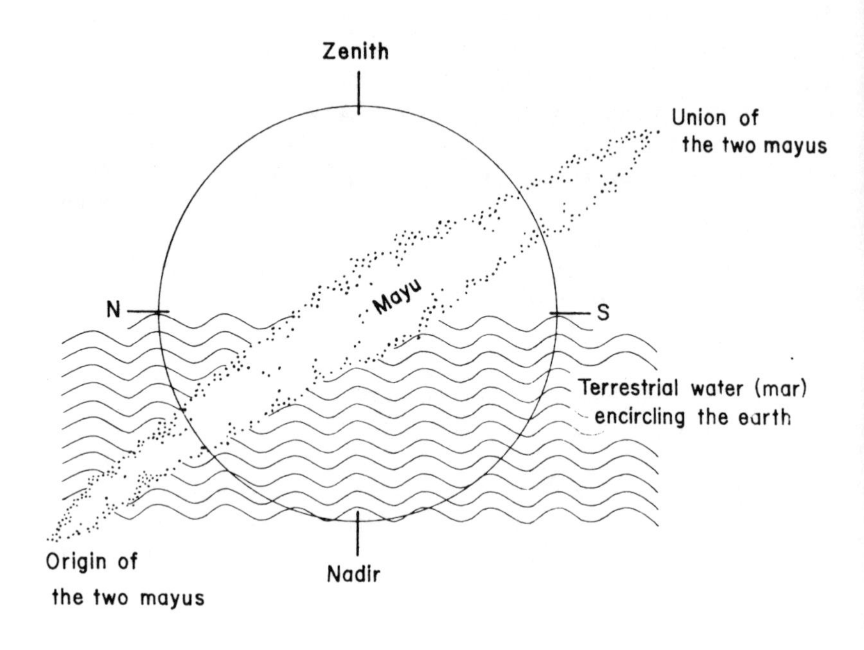

Fig. 20. The Cosmic Circulation of Water via the Milky Way

Thus, while terrestrial rivers conduct water downward (rain → streams → rivers → downward), the celestial River recycles water upward (cosmic ocean → northern Milky Way → upward). The Milky Way is therefore an integral part of the continual recycling of water throughout the Quechua universe.[4]

With this understanding of the cosmological significance of the celestial River in mind, we return to the problem of the quadripartition of the celestial sphere by way of the Milky Way. Study of the apparent rotation of the Milky Way around the earth reveals a pattern in which the southern and northern ends alternately (every twelve hours) rise from the southeast and northeast, respectively. Again, this alternation comes about because of the inclination of the plane of the Milky Way with respect to the plane of rotation of the earth's axis. Another observational phenomenon resulting from this inclination is that when either axis of the Milky Way passes through the zenith, the line which it forms across the sky will not be oriented straight north-south, which would be the case if it had the *same* plane of rotation as the axis of the earth. Instead, it will form two "axes" in the sky. One axis runs from the southeast through the zenith to the northwest; twelve hours later, when the other hemisphere of the Milky Way stands in the zenith, another axis will be formed which runs from the northeast through the zenith to the southwest. Thus, the overall, twenty-four-hour pattern of the Milky Way as it crosses the zenith is two intersecting, intercardinal axes. These relationships are diagramed in figure 21. In this way, the celestial sphere is divided into quarters (suyus).

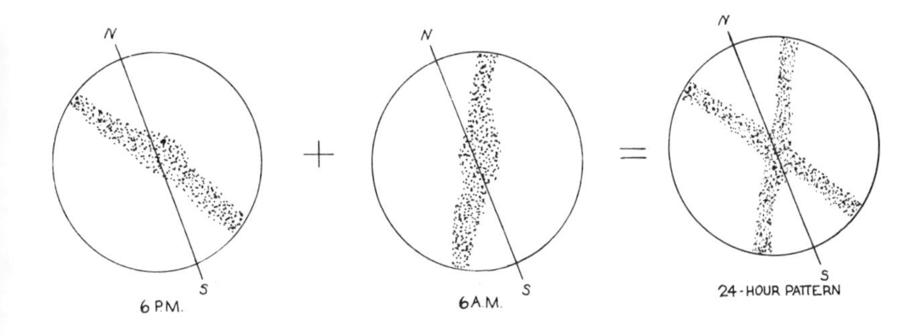

Fig. 21. Twelve-Hour and Twenty-Four-Hour Orientations of the Milky Way in the Zenith

In Misminay, the cross which is made in the zenith by the intersection of the two axes of the Milky Way is referred to as Cruz Calvario ("the Cross of Calvary"). As the two axes move either from the east to the zenith or from the zenith down to the west, they are referred to as *t'ihsu* ("tilted, tipped"). Recalling that the two terrestrial axes in Misminay, the footpaths, intersect at a point referred to as a cross (Crucero) and that the two axes of the quadripartition of terrestrial space are also intercardinal, we have two important clues for beginning to understand the similarity in the structure and organization of terrestrial and celestial space. Before elaborating upon this similarity, however, we must first consider some additional data concerning the Milky Way.

In a series of interviews with a paqo in Misminay, a man very skillful in the art of divination by means of maize kernels, I was told that the Mayu "comes from the sun" (*intimanta kashan*). This concept was confirmed by other informants, one of whom added that during the rainy season, from November through

February, the celestial River runs southeast to northwest and during the dry season, May through August, it runs northeast to southwest. This material is difficult to interpret, especially since the one informant was not interested in explaining how the Milky Way "comes from the sun," but it could possibly be explained in the following way.

When the Milky Way first becomes visible in the sky in the early evenings during the dry season, it stretches across the sky from the northeast to the southwest. However, during the rainy season, the Milky Way first appears in the evenings in a line running from the southeast to the northwest. It should now be recalled that on the solstice of June 21, which is about the middle of the dry season, the sun rises in the northeast; on the solstice of December 21, the middle of the rainy season, the sun rises in the southeast. (These seasonal orientations of the Milky Way and the rise of the solstices are represented in figure 22.)

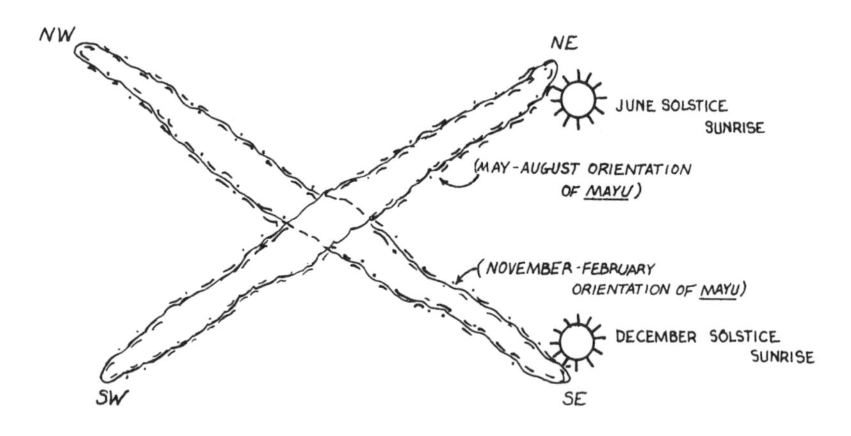

Fig. 22. The Solstices and the Seasonal Axes of the Milky Way

The celestial River, the Mayu, can be said to "come from the sun" in the sense that, during the dry season, they both rise from the northeast and, during the rainy season, they both rise from the southeast; that is, the seasonal changes in the early *evening* orientations of the Milky Way coincide with the annual changes in the early *morning* orientations of the sunrise. These seasonal coincidences are further related in that the periods around the two solstices are the only times when the sun actually passes through, and rises with, the Milky Way (see chap. 8 and Urton 1978b).

Correlating the Terrestrial and Celestial Quadripartitions. We now have a preliminary outline of the structural correlations between the organization of terrestrial and celestial space, central in which is the structural correlation, a kind of "mirror image," between the four quarters of the earth and the four quarters of the sky (see fig. 23). There is not sufficient data from Misminay concerning the shape of the underworld (otra nación) to extend the mirror image to that realm of the cosmos. However, at least one ethnographic account suggests that the cosmological structures illustrated in figure 23 may, in fact, be reflected into the underworld. The account comes from Fock's study of Juncal (n.d.), in the highlands of southern Ecuador. In Juncal, the earth (Cay-Pacha) is considered to be a plane encircled by the ocean (Mamacocha). Above the earth is the heaven or sky (Jawa-Pacha) in the shape of a dome; below is the underworld (Ucu-Pacha). Fock says that "*Cay-Pacha* and *Ucu-Pacha* are regarded as mirror-images" (p. 6).

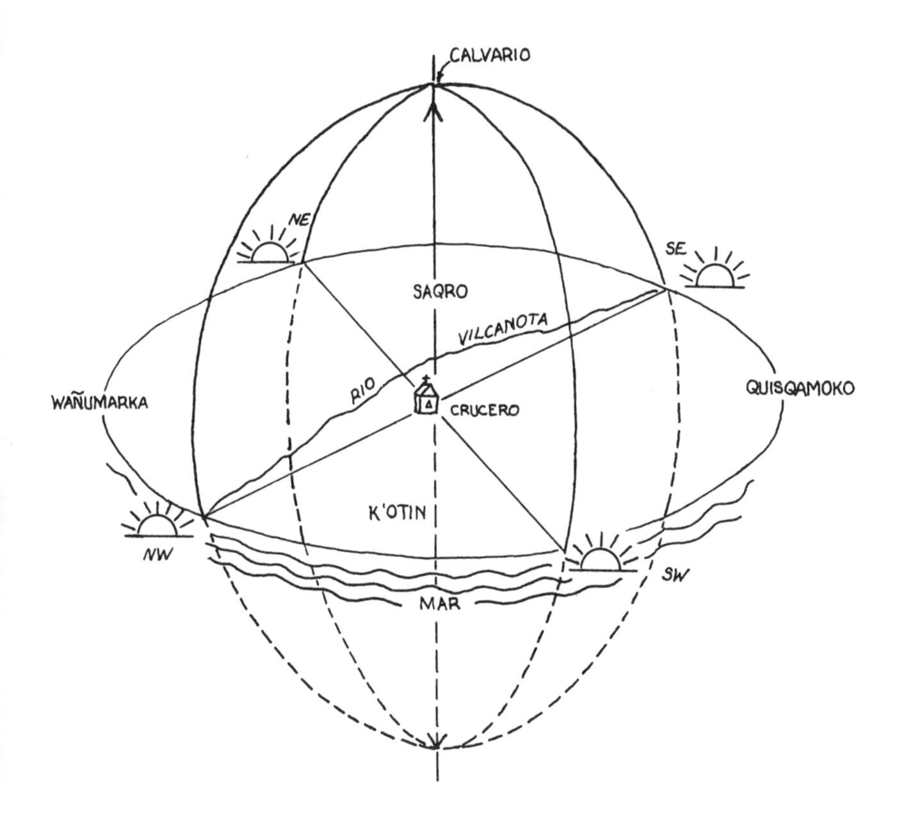

Fig. 23. The Cosmology of Misminay

One area for comparing the structural characteristics of the spheres of the universe in Misminay concerns the axes for the division of space on the earth and in the sky. To begin with, in both cases the divisions come about by the intersection of two intercardinal axes and in both cases the points of intersection are labeled "crosses" (earth cross = Crucero; sky cross = Calvario). The runa of Misminay state that Calvario is at a point straight overhead, in the zenith. Thus, the zenith cross articulates the two axes of the sky just as Crucero, the cross at the center of the community, articulates the two axes of the earth.

Second, the terrestrial and celestial axes are both associated with the movement of water. On the earth (i.e., in Misminay), water flows from the south-southwest to the north-northeast; in the sky, terrestrial water enters the celestial River in the north and flows southward. The water then returns to the earth in the form of rain, which is trapped in the reservoirs of Misminay and recirculated northward. Therefore, the two axes of the earth operate in conjunction with the two axes of the Milky Way to continuously circulate water throughout the universe. Since the water stored in the reservoirs is used to irrigate the maize and bean fields within Misminay, the existence of the community depends upon the continuous cosmic circulation of water.[5]

Third, though the orientation of local rivers and irrigation canals differs from community to community (resulting in differences in local cosmographies), for a large part of southern Peru around the Department of Cuzco the principal cosmic/terrestrial river is the Vilcanota (Urubamba) River. The Vilcanota, which is oriented southeast-northwest, is equated with the celestial River, the Milky Way. (Later we will see that the Vilcanota River also provides the course for the movement of the sun *underground* from *west to east* during the night.) An interesting datum supporting the correlation of the Vilcanota River and the Milky Way in other communities within the Department of Cuzco comes from the Province of Canchis. In divining for stolen articles, one observes the relationship between shooting stars and the Milky Way. If a shooting star moves toward the Milky Way, the thief is said to be going toward the Vilcanota River. The reverse is also true (Dr. Benjamin S. Orlove, personal communication, 1978). In addition, considerable ethnohistoric evidence affirms the cosmological significance of the southeast-northwest terrestrial axis of the Vilcanota River during Inca times (Zuidema 1978b).

Fourth, the north, the quarter of Apu Wañumarka, is associated with the concept of "origin." The origin of the celestial River(s) is in this quarter, as are a number of places on the earth associated with the ancestors of Misminay.

Fifth, we return to the nocturnal runa—whose primary orientation to the celestial sphere comes about by viewing the night sky in relation to named points on the horizon—and to the question "How can lines of sight be oriented once they leave the earth?" We may now conclude that the runa uses the alternating intercardinal axes of the Milky Way in conjunction with the zenith "cross" to orient his line of sight. This orientational process is similar to the way the house clusters and apus of the four quarters of the earth are oriented with respect to the two intercardinal terrestrial axes in conjunction with the center at Crucero.

Finally, I hypothesize that (a) the terrestrial axes are conceptually extended to the horizon and (b) the four points where the terrestrial axes intersect the horizon *ideally* coincide with the four solstice points. If these two hypotheses can be argued successfully, we will have a more solid basis for making a direct correlation between the structure of the terrestrial and celestial spheres.

3. The Sun and the Moon

Solar Cosmology and Astronomy in Misminay

Terminology and Symbolism. In addition to the frequent use of the Spanish term *sol* ("sun"), the most common name for the sun in Misminay is *inti*. These two terms, sol and inti, are used in most day-to-day contexts. However, I also heard the following names used on various occasions: *Nuestro Dios* ("our Lord"), *Otuño* ("autumn"), *Taytacha* (Jesus Christ and "[male] saint"), and *Huayna Capac* (the eleventh Inca king).

A small girl in Sonqo said that Otuño is the name of the person in the sun (*intipi*). Otuño, she said, is male. I did not hear this term used by anyone else in Sonqo or in Misminay. Another term heard only once was Nuestro Dios. An informant in Misminay said that the *compadrazgo* ceremony of cutting a child's hair for the first time should never be performed after sundown because it would displease *"inti, nuestro dios."* In Sonqo, a man whom Catherine Allen (personal communication, 1975) observed making a *tinkapu* (an offering of alcohol) to the setting sun alternately referred to the sun as "Jesús Cristo" and "Huayna Capac."

This equation of the sun with God/Christ is consistent with the designation of the sun as taytacha, a name normally reserved for Christ and, less often, for male saints. David Gow (1974:92 n. 23) reports that *taytacha* is also the general term used for crosses in the area of Ocongate. An informant in Misminay said that the sun is "taytacha" and that his name is "Huayna Capac." In this connection, I was told that even during the daytime the stars are in the sky; they disappear every day because of the appearance of Huayna Capac (the sun). Juvenal Casaverde Rojas (1970:167) reports from the community of Kuyo Grande that the sun is considered to be male and that he is the son of the Virgin Mary; his name is Huayna Capac. In Chinchero (J. Núñez del Prado 1970:

95) and Kauri (Mishkin 1940:235), the sun is associated both with Manco Capac, the first Inca, and with Huayna Capac, the eleventh Inca. Thus, there is a syncretism of the Spanish Catholic concepts of dios and taytacha, symbolized by the cross, with the king(s) of the Inca empire.

Reckoning Daily Time by the Sun. The sun is used for coordinating activities, for describing events in the past, and as a way of timing appointments in the future. It is said to move through the sky *como un reloj* ("like a clock"); it rises in the east about 6:00 A.M., stands over head at noon, and goes into the mar (cosmic sea) in the west at about 6:00 P.M. In most day-to-day references of this type, the sun is spoken of as "inti"; I have never heard the name Huayna Capac applied in such contexts. Quechua numbers are sometimes used in stating the time of day, but Spanish numbers are much more common. When arranging to meet someone the next day, for instance, the time is agreed upon by agreeing on the place in the sky where the sun will stand when the meeting is to take place. On such occasions, I have noticed that the position of the sun is indicated not just for the altitude of the sun along an east-west line, but also that its approximate north-south declination is indicated.

In an irrigation-canal-cleaning ceremony which took place in Misminay on April 29, 1977, the end of work and the beginning of a feast was determined by the sunset. Work ceased about thirty minutes before sunset, and the participants sat quietly drinking chicha in a semicircle open to the east. At the moment of sunset (lower tangent), everyone turned to look at the sun, dropped to their knees, and a prayer was said in Quechua. The timing of rituals and ceremonial activities by the sun has also been reported in the pilgrimage to the sanctuary of Coyllur Riti at the time of Corpus Christi (Sallnow 1974:114-116).

The Sun in the Cosmology of Misminay. Each morning, the sun rises in the general direction called *inti seqamuna* ("place from where the sun rises"), the only Quechua "directional" term I encountered. *Mar*, the Spanish word for *sea,* is often used to refer to the west, but it is also commonly used for the north, south, and west in opposition to inti seqamuna, the east. An integral part of mar, the cosmic sea or ocean surrounding the earth, is the Vilcanota (Urubamba) River, which communicates between the mar in the east and the west. The valley of the Vilcanota River makes up part of the view from Misminay toward the north-northeast. The Vilcanota, running from the southeast to the northwest, is the terrestrial reflection of the path of the sun through the sky during the day and is considered the actual path of the west-to-east movement of the sun during the night. According to an informant in Misminay, "The sun rises from the east and moves through the sky from east to west. When it sets in the west, it enters the "sea" or "other world" [*otra nación*]. After entering the sea or the other world, the sun makes a twisting motion to the right [north] and begins its journey back to the east *beneath the Vilcanota River.* It takes all night for the sun to move from the mar to inti seqamuna."

The nightly union of the sun and the Vilcanota River is important for understanding the structure and operation of the cosmological system in Quechua communities, but it is perhaps more important as an expression of the circum-stances that are believed to account for differences in the power and size of the sun from the rainy season to the dry season. During the rainy season (November-February), the sun is larger, brighter, and hotter because, while traveling under the river all night, he drinks from the swollen waters of the Vilcanota; thus, he is very powerful when he rises in the morning. However, during the dry season (May-August), the Vilcanota carries considerably less water, and the sun is weaker when he rises because he has had less to drink during the night.

In this way, the Vilcanota River and the sun are locked in a pattern of daily interaction which, over the course of a year, reflects the cyclical change from rainy season to dry season. The coldness of the dry season (= winter) is a result of the smallness of the sun, which is a result of the small amoung of water in the Vilcanota River. The opposite, as we saw, occurs in the summer. Casaverde Rojas 1970:167) describes a similar pattern of solar cosmology in the community of Kuyo Grande, where during the night the sun is said to move from the west to the east via a subterranean tunnel. The tunnel is filled with the waters of the *mar qocha* ("sea lake"); the sun drinks from these waters every night.

A few additional observations should be made at this point. First, the Vilcanota and the sun do not interact in isolation. The rhythmical (seasonal) swelling and diminishing of the Vilcanota River is related to the swelling and diminishing of the smaller tributaries above the main course of the river. These local river and *acequia* (irrigation canal) systems, such as the one described for Misminay in the previous chapter, ultimately account for variations in the amount of water in the Vilcanota. Therefore, the local Misminay hydraulic system operates in conjunction with the Milky Way, which is equated with the Vilcanota River, to circulate water continuously throughout the terrestrial and celestial spheres. The Vilcanota/sun system must thus be seen as one part of a larger, cosmic system.

Second, the seasonal cycling which results from the interaction of the sun and the Vilcanota River is related to the seasonal variations in the orientation of the sun and the celestial River, the Milky Way (see fig. 22). Finally, it should be noted in passing that certain characteristics of solar haloes are believed to vary with the seasons. Rainy-season solar haloes are typically of three certain colors; solar haloes of the dry season exhibit three different colors (see chap. 4). Because observing solar haloes figures in the prediction of rain, the haloes are also involved in the larger cosmic system which accounts for the circulation of water (energy) throughout the universe. This larger system includes rain, terrestrial rivers, the Vilcanota as a "cosmic river," the sun, and the Milky Way.

The Solstice Sun. The (apparent) annual north-to-south-to-north movement of the sun, in almost all societies, is minimally divided into two parts determined by the points at which the northward and southward motions of the sun are reversed. These points of reversal are referred to in Western astronomy as the solstices. In Misminay, as in most contemporary Andean communities, the celebrations of the solstices are syncretized with the Catholic celebrations of Navidad, December 25, and San Juan, June 24 (cf. Foster 1960:207-208).

One of the most interesting associations of the December solstice is with celestial and terrestrial foxes. The Atoq ("Fox") is a rather amorphous dark-cloud constellation which stretches at a right angle from the tail of Scorpio crossing the ecliptic between Scorpio and Sagittarius (# 35 in the Star Catalogue of table 7 [chap. 5], hereafter referred to as *SC*). One informant referred specifically to this relationship between the Fox and the ecliptic by mentioning that, at times, the moon rises with the celestial Fox. In addition to this explicit association of the moon and the Fox, a number of informants indirectly pointed out a similar association between the sun and the Fox. However, in order to understand the full calendrical and astronomical significance of the solar-Fox relationship, we must first analyze the various contexts in which it is found.

While discussing the stars with one informant in Misminay, I was asked to name some of the constellations I had found so far in the community. I began naming various star-to-star and dark cloud constellations. When I mentioned the word *atoq* ("fox"), the informant, with a spontaneity similar to free association, said "Wañumarka," which means "storehouse of the dead" and is the name of the principal apu of Misminay. On another occasion, as I was walking up Wañumarka with a young man from Mistirakai, he began talking about the contors, atoqs, and *zorritos* ("little foxes," i.e., skunks) that live on Wañumarka; they are found especially on the Huaylanka side of Wañumarka, he said. The June solstice sun (which sets at an azimuth of about 296°) sets between Cerro Huaylanka and Apu Wañumarka (see fig. 14). Finally, on New Year's Eve, 1976, I was told that Apu Wañumarka is always full of foxes at this time of year because baby foxes are born there every *December 25,* (four days after the solstice). My informant cautioned that it would be dangerous to go alone to Wañumarka at this time.

Now, if we consult a celestial globe, we find that the sun rises into the dark cloud constellation of the Fox, and coincidentally into the Milky Way, from about December 15 to December 23. This means that at the same time that the December solstice sun rises with the celestial Fox in the *southeast,* terrestrial foxes are born on Apu Wañumarka in the *northwest*, in the *antisolstitial* (the June solstice sunset) direction. In effect, this results in a solstitial/antisolstitial (December solstice rise/June solstice set) axis, which is timed by the

December solstice passage of the sun through the Milky Way and the dark cloud constellation of the Fox.

The southeast-northwest intercardinal (solstitial) "Fox axis" is close to the southeast-northwest axes of the Vilcanota River and of the Milky Way during the rainy season (November-February). In discussing data on the Pleiades (the celestial Storehouse), we will find that this *December solstice, southeast-northwest, Fox axis* has a complement in a *June solstice, northeast-southwest, Storehouse axis,* which is timed by the rise of the June solstice sun with the Pleiades in the northeast in opposition to the setting of the tail of Scorpio in the southwest. Besides demonstrating how specific points of terrestrial and celestial space are integrated, the example of the Fox axis suggests that myths about foxes and about foxes and contors (e.g., Ravines 1963-1964) can be sources for analyzing the Andean cosmological and calendrical tradition.

The Three "Sections" of the Sun. In addition to the solstitial division of the year into two halves, the space traversed by the sun throughout the year *may* be further divided into any number of subunits. These further divisions, however, are purely cultural constructs in that they are not based upon observable "facts" such as the solstitial change in the direction of the motion of the sun. In Western astronomy, and in most astronomical systems of the northern hemisphere, the equinox (the point midway between the two solstices) has been assigned tremendous cosmological importance as one of the two axes for the cardinal division of space into four parts; cardinal east and west are based, respectively, on the equinox sunrise and sunset.

This introduction is necessary because in Misminay the Spanish terms for east and west are known and used, but one soon realizes that *este* and *oriente* ("east") are not equivalent to cardinal east. Instead, they refer to the direction from Misminay to the distant-yet-visible community of Chinchero, which is at an azimuth of about 105° (15° south of cardinal east). What accounts for this "disorientation" of east?

In Misminay, the celestial and horizon space traversed by the sun during the course of a year is divided into three *partes* ("parts" or "sections"). These solar sections are labeled *A, B,* and *C* in figure 24, which includes the azimuths, when determinable, to the six places that are considered to form the boundaries of the sections of the sun as viewed from Misminay. The lower section, the area below the Cuzco-Ancascocha line, is not labeled because the sun does not rise south of the azimuth of Cuzco; that is, south of an azimuth of 125°. However, the sun does rise north of Calca (azimuth 80°). From Misminay, the June-solstice sun rises at an azimuth of 62°. Thus, section A contains some 18° of space along the eastern horizon which is covered by the sun at the time of its northern extreme (the June solstice).

The informant who supplied these particular data on the three sections of the sun also said that the sun is *en el centro* ("in the center") at noon on the day that it rises at Chinchero and sets at Cerro Huaylanka. These and other accounts relating to the "center" sun will be discussed shortly, but it is first necessary to analyze the division of the solar territory into three sections.

In collecting data on the sun, I was especially interested in knowing whether or not a "horizon calendar" is used to determine such events as when to plant and harvest and when to celebrate festivals. If the data I collected on this subject (see table 5) are applied to the solar sections (fig. 25), we find that section A is related to the area of sunrise at the time of the harvest (May to early July) and section B is related to the southward movement of the sun during the period of planting. If sections A and B are used for indicating those areas along the horizon where the sun rises at two important periods in the agricultural cycle, then what is the function of section C? In discussing the movement of the sun through the sky at different times of the year, an informant began by describing its movement from the northern (June) solstice southward to the zenith (which at the latitude of Misminay, occurs on October 30 and February 13). However, when I asked him to continue, to tell me about the sun when it is south of the zenith, he waved off the question, indicating that that part of the sun's move-

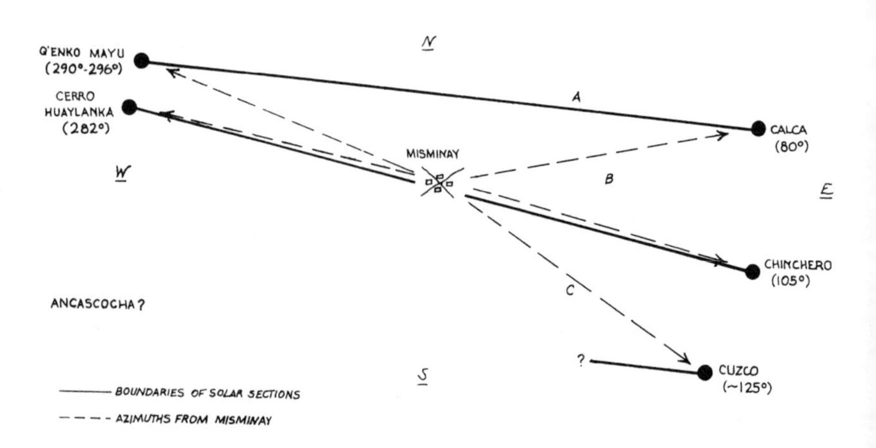

Fig. 24. The Sections of the Sun

Table 5. The Horizon Calendar

Event	Period	Azimuth (place of sunrise)
maize planting	August	85° (Paucartambo)
maize planting	August-September	80° (Calca)
maize planting	September-October	105° (Chinchero)
potato planting	August-September-October	80° (Calca)
maize harvest	May	70° (mtns. above Calca)
potato harvest	May	67° (mtns. above Calca)

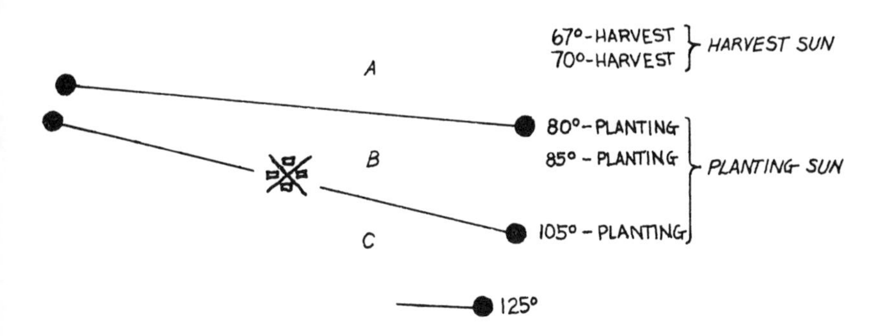

Fig. 25. The Sections of the Sun and the Agricultural Cycle

ment (or the sky in that direction) is uninteresting or unimportant. The informant's attitude concerning this portion of solar territory gives a clue to the purpose, and celestial associations, of section C. The purpose is related to the fact that the intense period of work during the planting season, which I will refer to as B_1 (B_1 = the sun moving *southward* through section B) is followed by a long period when the major "work" is the growth of the crops. After the crops are planted, it is no longer necessary to observe the sunrise or sunset in order to time agricultural duties. There are, of course, a number of tasks to perform during this period (see fig. 1), but these are secondary in a way; the full attention of the community is focused on the crops and bringing them to a succesful harvest.

The earth (Pachamama) is considered to be female in Quechua cosmological symbolism (Gow 1976a:209, Isbell 1976:55, J. Núñez del Prado 1970:72-73, and Wagner 1978:50). When the crops are inside the female earth, they are in the domain of the moon, a celestial body classified as female (Mama Quilla = "mother moon"). Thus, during the period of the sun's movement from the zenith to the south and back, the principal celestial body is the moon, not the sun. This interpretation is consistent with studies dealing with the symbolism (e.g., Brownrigg 1973:53), especially the sexual symbolism, of rituals which occur during the rainy season (the rainy season = the growing season). In fact, the indigenous chronicler of Inca culture Felipe Guamán Poma de Ayala (1936:ff. 894-895) tells us explicitly that in the Inca calendar the moon "governed" the period from August to December.

In the solar calendar of Misminay the pivotal direction or point on the horizon is Calca, located at an azimuth of 80° (10° north of the equinox). When the sun rises from Calca in late August on its way south, it is time to plant; when it rises from Calca in April on its way north, it is time to begin the harvest. Thus, the three "sections" of the sun are related to the principal periods of the agricultural cycle in the following manner:

Section B_1. August-October = planting (dry/rainy)
Section $C + B_2$. November-April = growth (rainy)
Section A. May-July = harvest (dry)

These associations are diagramed in figure 26.

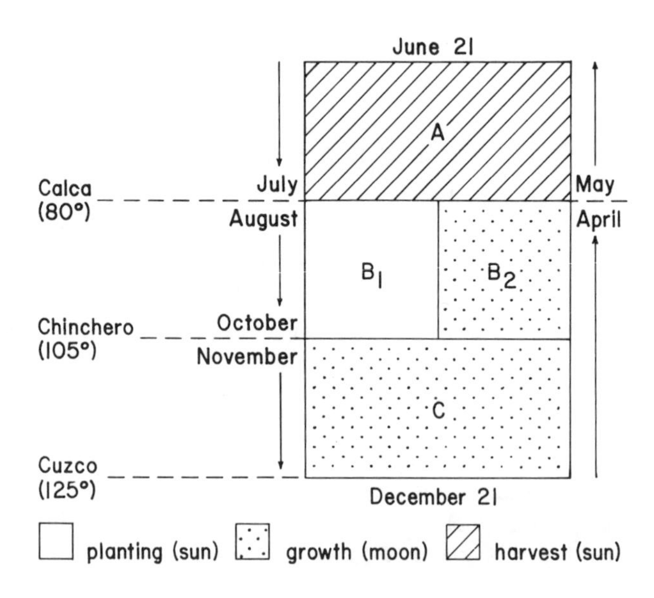

Fig. 26. The Calendrical Periods of the Sections of the Sun

The Center Sun. The informant who described the three sections of the sun said that, when the sun rises from Chinchero and sets behind Cerro Huaylanka, it passes through the "center" at noon. This is but one of several accounts I collected concerning the rise and set points of what I assumed to be the zenith sun, the sun which stands directly overhead at noon. At the time of my fieldwork, I was struck by the confusion among the different accounts. The Misminay runa seemed familiar with the concept of a "center" sun, but their descriptions of it presented many contradictions and inconsistencies.

My own misunderstanding stemmed from the knowledge that, since Misminay is located at a latitude of $-13°30'$, the sun passes through the zenith at noon on the day that it rises and sets about 13°30′ south of the east-west line—that is, at an azimuth of rise of 103°30′ and an azimuth of set of 256°30′. I also knew that the sun rises and sets at these points twice a year: on October 30 and again on February 13.

The "inconsistencies" in the various accounts are evident in table 6, in which I have listed the major accounts of the rise and set points of the "sun in the center," and in figure 27, which is a visual representation of the accounts.

Table 6. Azimuths of the Center Sun

Account	Rise		Set	
	Place	*Azimuth*	*Place*	*Azimuth*
1	Calca	80°	Apu K'otin	ca. 250°-255°
2	Chinchero	105°	Huaylanka	ca. 281°
3	Calca/Chinchero	80°/105°		
4	Chinchero	105°	Apu K'otin	ca. 250°-255°

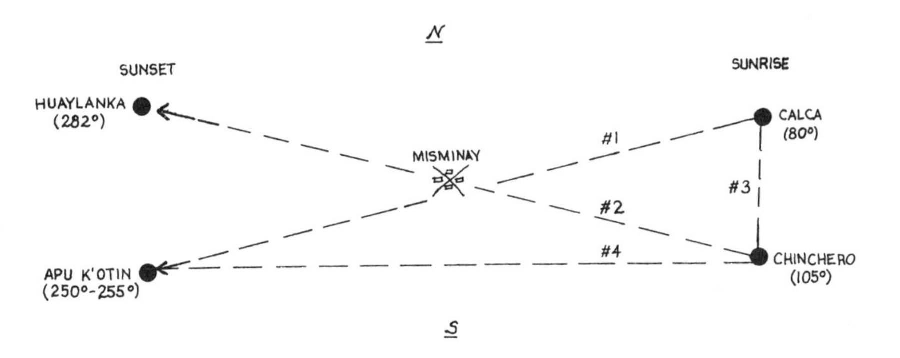

Fig. 27. The Center Sun from Misminay

Account 4 very nearly gives the correct azimuth values of 103°30′ and 256°30′ for the zenith sunrise and sunset. But why the confusion in the other accounts? And why especially the problem in account 3, in which the informant was unable to decide whether the center sun rises 10° north of east or 15° south of east?

These data can best be understood in relation to the three sections of the sun and to the agricultural periods with which they are associated. In Western astronomy the sun in the "center" refers to the zenith sun, but in Quechua astronomy it refers instead to the sun in the center *section*. That section, the "planting sun," covers the area of the horizon from azimuth 80° to azimuth 105° (see fig. 25).

The "planting sun," section B_1, is the only one of the three sections whose limits must be fixed fairly precisely because it is concerned with the determination of the time to plant maize in relation to the period of winter frosts. While traversing the center planting section, the sun rises in August in the northeast, from above Calca, and then moves steadily southward until it rises at Chinchero in late October and early November. August is at the end of the Andean winter and the beginning of spring. Therefore, crops (especially maize) planted *before August*, before the sun rises at Calca, are in danger of being destroyed by late winter frosts. However, since the growing season of maize at the altitude of Misminay is around seven to eight months, crops planted *after October* may not have time to mature before the frosts begin the next winter (in June). Therefore, the limits of the center-sun/planting-sun in Misminay are calculated fairly precisely to provide for the survival of the crops.

It is also important to note that Calca and Huaylanka, both of which are located about 11-12° *north* of the east-west line as viewed from Misminay (see fig. 14), are related to the concept of the sun in the "center." While the sun stands in the zenith at noon on the day that it rises 13°30′ *south* of the east-west line, it will stand in the *nadir at midnight* when it rises 13°30′ *north* of the east-west line. R. T. Zuidema (n.d.c) has shown that the antizenith (the nadir) line was used to orient certain of the lines of the ceque system of Cuzco during Inca times. This same concept may be reflected in these data from Misminay; that is, the limits of the "center sun" may correspond to the four points of the zenith and nadir sunrise and sunset. It should also be pointed out that the *full moon* stands in the zenith at midnight only when the sun is at the nadir at midnight. Therefore, the observation of the nadir sun can be performed, within a small variation determined by the 5° inclination of the lunar orbit (Aveni 1972:532), by observing the full moon in the zenith at midnight.

This is not the first time that we have encountered a method for timing the planting season in Misminay, for we found that the four saints' days, celebrated in Misminay from July 26 to October 4, span the planting period (fig. 3). The concept of a center/planting sun is a pre-Spanish calendrical construction, and we can hypothesize that, in Misminay at least, the pre-Spanish center/planting sun has been syncretized with a similar period of time in the Catholic saints' day calendar. The correspondence between these two periods of time can be seen in figure 28. It is clear from this figure that although the correspondences between the saints' days and the temporal borders of the center/planting sun are not exact, they are near enough to allow us to hypothesize that the festivals may all be related to Incaic planting festivals and perhaps derived from them.

As for the assertion that this period of time and solar space was important during pre-Spanish times, we have the testimony of the Anonymous Chronicler (1906). In discussing the pillars which the Inca erected in Cuzco in order to observe the movement of the sun, he noted the following:

Along the high ridge of mountains to the west, in view of the city of Cuzco, they placed four pillars in the form of towers at a distance of two or three leagues from Cuzco; in place, two hundred paces separated the first pillar from the last; there were 50 paces from one of the middle pillars to the other; and the two extreme markers fit the calculated limits; so that when the sun entered the first pillar, it was a warning for the general period of planting, and they began to plant crops in the fields higher up . . .; and when the sun entered between the two central pillars, it was the period and general time for planting in Cuzco, *this was always in the month of August* (Anonymous 1906:151; my translation and emphasis).

It seems reasonable to conclude that the concept of the center/planting sun (the sun from August to late October) which is found today in Misminay is a "descendant" of the Incaic practice of calculating the time of planting maize by means of solar pillars. Even though the solar pillars (sucancas) which were observed in Cuzco during the time of the Incas were systematically destroyed during the "extirpation of idolatries'/ (Arriaga 1920), we may see the virtual embodiment of them in the saints' days (i.e., four saints' days = four solar pillars) and in such concepts as a planting and harvesting sun.

Lunar Cosmology and Astronomy in Misminay

In comparison with the data on the sun, I was able to collect less specific information relating to the moon. The reason for this is twofold. First, the different cycles of the moon make it a much more complicated celestial body to observe and discuss than the sun. Its north-south-north sidereal movement along the horizon is greater by a few degrees than that of the sun, and it goes through more than thirteen of these sidereal cycles during the period of one solar year

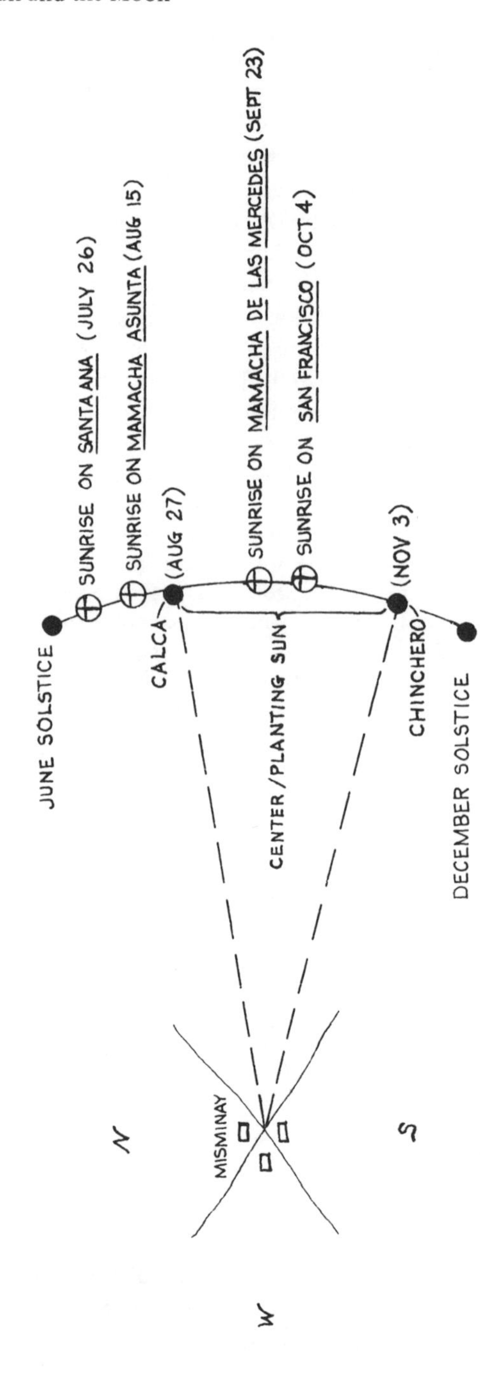

Fig. 28. The Saints' Days and the Center/Planting Sun

(thirteen sidereal lunar months of 27.3 days = 354.9 days). In addition, the moon has another periodicity, the 29.5-day cycle of the lunar phases (the synodic lunar cycle). The moon goes through more than twelve synodic periods in the course of a solar year (twelve synodic lunar months of 29.5 days = 354 days). Even though thirteen sidereal months are close to twelve synodic months, they are still only "close" and both periods are well short of the 365¼-day solar year. This difference, as any student of ancient and primitive calendar systems knows, is the correlation problem which is the Achilles' heel of every attempt at the calendrical correlation of the solar and lunar cycles.

However, the principal reason for my own difficulty in collecting lunar data is related to the fact that the moon is female in Quechua astronomical symbolism; her periodicities are closely tied to the periodicities of the female human body, especially the menstrual cycle, and it is difficult for a male anthropologist to pry very deeply into this area of astronomical knowledge and cosmological symbolism. It is an area of ethnoastronomical research, I should think, which can be thoroughly studied only by a woman.

The association of women and the moon is, of course, neither unique to the Andes nor is it a totally arbitrary association. The female menstrual cycle has a periodicity which falls more or less between the sidereal and the synodic lunar cycles; thus, it would be more surprising *not* to find a female/lunar association in a particular astronomical system. However, it is not generally recognized that the importance of the moon for females also involves the stars. In correlating the menstrual and lunar cycles, it would become a regular practice to observe not only the phases of the moon, but also its movement against the background of the stars. These periodic observations could develop naturally into private lunar zodiacs. Collectively, this process could easily lead to the standardization of a female lunar zodiac. This hypothetical process of evolution from uncoordinated female-lunar zodiacs to a standardized zodiac is not completely lacking in a biological (menstrual) basis. It has been shown that when women live together in isolation over long periods of time, their menstrual cycles tend to become synchronized. It is interesting to consider this fact in relation to the institution of the Acllas, the "Virgins of the Sun," in Incaic society. The Acllas lived together in a palace in Cuzco called the Acllahuasi; their lives were dedicated to the performance of religious rituals and ceremonies in the Temple of the Sun (Coricancha). Thus, the Acllas could well have served not only as the "biological standard" for a coordinated lunar zodiac, but also as the record-keepers of nocturnal celestial cycles.

In discussing the stars and constellations with female informants, I often found that they tended to rely on the moon for providing a point of stellar orientation. It was especially interesting to find that when locating constellations on a star map, for instance, women were almost always disoriented by the absence of a moon on the map. It was common to have women look at a star map and immediately ask where the moon was. In contrast, the absence of a moon did not seem to present a problem to male informants.

Terminology and Symbolism. The most common name for the moon is *quilla* ("moon," "month"). However, just as the sun has one name (inti) which is used in most day-to-day contexts and others which are used for the sun as a male (e.g., Huayna Capac), so too does the moon have female names. The following names were used in references to the moon by different informants: *Mama Quilla* ("Mother Moon," "Mother Month"), *Mayor* or *Warmi Mayor* ("Principal," "Principal Wife/Concubine"), *mamacha* (The Virgin Mary and "[female] saint"), and *Coya (Colla) Capac* ("Noble or 'principal' Queen.")

The name Colla Capac was obtained in a situation which deserves some elaboration since it demonstrates the flexibility of astronomical terminology and symbolism. During an interview with a paqo in Misminay, the following people besides me were present: my principal informant (a young man of 25), his wife, his grandfather (the paqo), and his grandmother. At the end of the interview, we began discussing the name and sex of different astronomical phenomena. Everyone agreed with the grandfather that the sun is taytacha (Christ; "male saint") and that his name is Huayna Capac. I then asked about the moon, and the grandfather answered very authoritatively that "he" is also taytacha. At this, the women objected immediately and vociferously. The moon, said the grandmother, is most certainly *not* taytacha, *she* is mamacha. The grandfather became very flustered and looked as though this complication was calling his whole credibility into question. In order to salvage the situation, he admitted that yes, the moon is indeed mamacha, and that her name is Colla Capac. The women seemed somewhat less than satisfied with the name (no female informant independently gave this name for the moon) but to have objected further would have been a real embarrassment to the grandfather. Besides, the main point had been made; the moon is female, *not* male.

I think that some of the lingering reserve on the part of the women was due to the fact that the name Colla Capac for the moon placed her in a dependent or subservient status to the sun, Huayna Capac. The name *colla* ("queen") indicates the royal status of the moon. However, in a lunar context, the term appears to be related to Qholla. In the biological metaphor applied to the lunar phases, Qholla refers to the moon when it is just born. The moon is perhaps related to the sun, even by as intimate a relationship as wife or sister, but she nonetheless retains a degree of independence within the domain of woman and should not, properly, be classified as the "child" of the sun. In other words, the grandfather was speaking out of turn, and though he might be allowed to give a male-oriented name for the moon, he could not be allowed to change her sex altogether.

Bernard Mishkin also found the use of the name Colla Capac for the moon in the community of Kauri. In Kauri, the sun (= Huayna Capac and Manco Capac) is related to agriculture and the moon (Colla Capac) is related to pastoralism (Mishkin 1940:235-236). As in Misminay, Juan Núñez del Prado (1970: 95-96) found in Qotobamba that the moon is called Mama Killa; she is a "beautiful woman with a white hat." In Kuyo Grande, the Virgin Mary is identified either with Pachamama (the benevolent female deity of the earth) or with Pacha-Killa ("earth/time moon"; see O. Núñez del Prado 1973:40). Juvenal Casaverde Rojas (1970:168) also reports the name Killa Mamacha for the moon in Kuyo Grande; she is considered to be the wife of the sun. The belief that the moon is the wife of the sun may also be reflected in the name Warmi Mayor found in Sonqo. *Warmi* means "wife" or "concubine" (see González 1952).

The spots in the moon are also important named celestial objects. One informant, a young girl from Sonqo, said that the spots represent an *urqu*, a word which can be translated as "mountain" but which is more commonly used for male llamas and for males in general. This identification comes from the same informant who called the woman in the moon Warmi Mayor; thus, she couples the woman in the moon with a male llama. The association of a woman and a llama with the lunar spots has further support.

Two of my principal informants in Misminay said that mamacha is in the moon and that during the full moon she can be seen holding her baby daughter. At other times, however, mamacha is seen with a *caballo* ("horse"). Another informant from Misminay said that the figure in the full moon is mamacha riding a horse; the horse is pawing at the ground with its front hoof. Thus, there seems to be the fairly consistent association of an important woman and an animal (llama/horse) with the spots in the moon.[1]

The identification of the dark lunar spots with a llama is interesting because one of the prominent dark cloud constellations is also identified as a llama (*SC* # 40). In addition, beneath the dark cloud constellation of the Llama is a smaller dark streak referred to as *uñallamacha* ("suckling baby llama"; *SC*# 45). The thin crescent of the new moon is sometimes referred to as *uña quilla* ("suckling moon"). Therefore, we find the following llama/moon relationships:

first crescent moon and baby llama constellation = uña ("suckling . . . .")
spots in the full moon = urqu (male llama)
dark-cloud constellation = Llama (and Mamacha?)

The equation of the lunar spots with the dark clouds in the Milky Way was also current in fifteenth-century European astronomy. For example, Gaetano de Thiene (1387-1462) "compared the lunar spots with the shade of the Milky Way, arguing that in both cases the physical cause was a higher density of the ethereal material" (Jaki 1972:49-50). These associations must be further investigated in the field in order to determine more precisely the nature of the symbolic relationships and astronomical periodicities of the lunar and stellar llamas (see also Zuidema and Urton 1976).

The Synodic Lunar Cycle. In Quechua astronomy, the phases of the moon are important for timing various activities. On several occasions, I asked informants to supply names for the full, half, and new phases of the moon; figure 29 indicates the terms most frequently given.

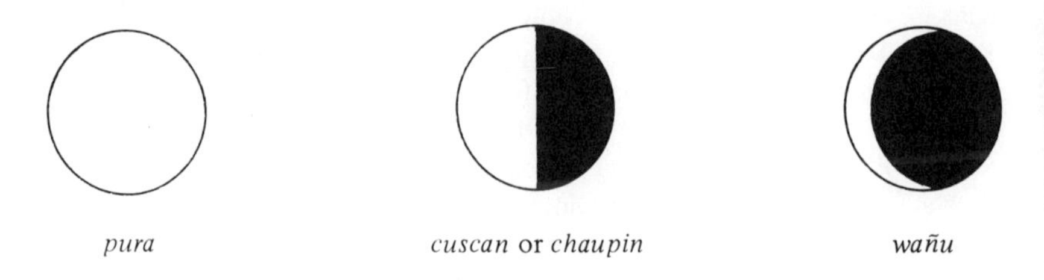

Fig. 29. Lunar Synodic Terminology

These terms, especially the ones for the full and new moon, are similar to the lunar-phase names mentioned by other ethnographers. For instance, J. Núñez del Prado (1970:96) and Casaverde Rojas (1970:168) both discuss *pura* and *wañu* as though they are separate phases of the moon. This was my own impression early in my fieldwork, and it was only after an informant from Mistirakai became immensely frustrated with my Western way of thinking of the lunar phases (i.e., of labeling pura and wañu as separate, fixed lunar phases) that he gave the following explanation and demonstration.

He first picked up two round cow chips and superimposed one upon the other; one complete cow-chip circle was invisible behind the other. The one in front he called pura, the one behind, wañu. Then, he began to expose more and more of the wañu cow chip, finally superimposing wañu on top of pura; pura was now invisible (see fig. 30). There are, he said, two parts of the moon: wañu and pura. One may see a complete circle of either one of them but the other *complete* circle will always be present behind or within the visible one.

As mentioned earlier, the first crescent new moon is sometimes referred to as *uña* ("suckling"; see Lira 1946:18). There is a suggestion in this terminology that the sequence of lunar phases can also be likened to the states of human maturation. In the area around Ocongate Percy Paz Flores (personal communication to Dr. R. T. Zuidema, 1976) collected names for the lunar phases that support this notion (see fig. 31).

From these data, we again see that pura is not thought of as a fixed, separate phase of the moon; rather, it is one stage in a sequence built upon a biological

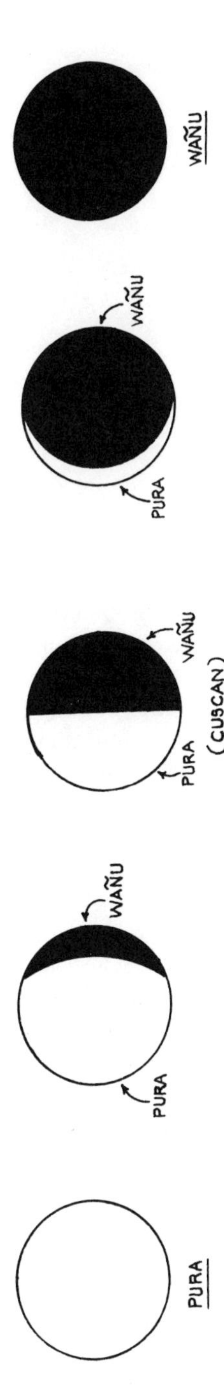

Fig. 30. The *Pura/Wañu* Lunar Sequence

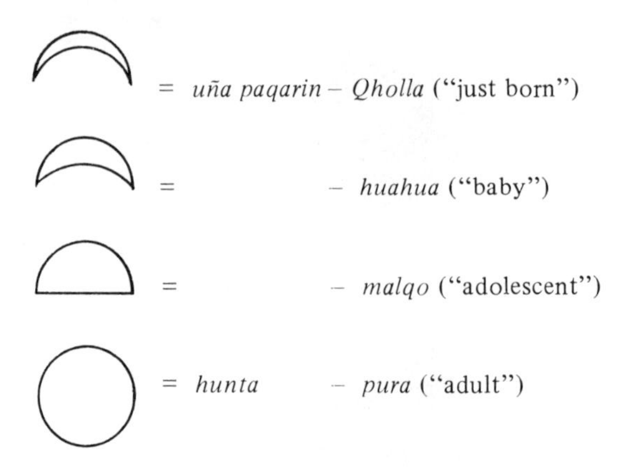

Fig. 31. Lunar Terminology in Ocongate

metaphor. The metaphor will be understood better if we examine the three basic terms I used in figure 30 to describe the synodic lunar sequence: wañu, cuscan, and pura.

a. *Wañu*. The best translation for this term is "inanimate" or "lifeless"; as such, wañu defines one end of the continuum animate/inanimate, or life/lifeless. It does not, however, refer to an absolute or static point in the sequence alive/dead.

b. *Cuscan*. This term refers to the union of two equal parts. For instance, in the continuum animate/inanimate, cuscan is the point at which an animate unit is in opposition to an equal inanimate unit. It must be emphasized that cuscan contains no indication of the direction which future changes between the two equal units might take; they are at a state of equilibrium in time and space; subsequent motion may proceed in either (or any) direction.

c. *Pura*. This term refers to a concept of classification, rather than to a specific object or action. As a classifier, pura groups together all the members of a class or sequence which are in a *relationship of interaction.* In the sequence inanimate/animate, where wañu stands at one end representing inanimate, pura stands at the other end representing animate *in relation to* varying degrees of inanimate. Thus, wañu ("inanimate") is the point of orientation for the synodic lunar cycle.

This analysis of the terminology of the lunar phases clarifies the use of the biological metaphor suggested by the data in figure 31: humans first do not exist, then they exist, and then they exist no longer. The fundamental sequence of life has its analogy in the moon, which is first inanimate (wañu), then goes through several stages of animacy (uña, pura, and so forth) and then returns to inanimacy (wañu).

In order to understand the full significance of the lunar biological metaphor, we need to investigate more fully the dark and light synodic cycle of the moon and its symbolic relationship to the menstrual cycle as the ultimate expression of the human cycle of animacy/inanimacy. Sexual and menstrual symbolism are also found as characteristics of beliefs about the moon in other South American Indian groups. In northwestern Brazil, the Surára and Pakídai Indians imagine that the moon consists of two half-moons held together by a river of blood which flows north-south. The eastern half of the moon, the bright half, is male; the western dark half is female (Becher 1976:338-341). The dark lunar spots are themselves often associated with menstrual blood (see Lévi-Strauss 1978: 291-292 and Reichel-Dolmatoff 1975:137).

The Moon and Agriculture. In Yucay, I was told that the planting of maize should begin with the new moon and extend to the full moon, by which time planting should be completed. In 1975, the maize planting in Yucay did in fact begin with the new moon, but it extended about a week beyond the full moon. My impression is that the planting of maize should be done, in general, during the period of the waxing moon (decreasing wañu).

In Misminay, it is said that only the planting of potatoes is timed by the moon. Potatoes are thought to produce best when planted during the waning moon (increasing wañu). Although informants in Misminay maintained that the lunar phases are unimportant for planting crops other than potatoes, I would suggest the following general relationships between the planting of crops and the lunar phases:

a. Pura → wañu ("waning"). Plant crops which produce underground (e.g., potatoes, oca, ullucu).
b. Wañu → pura ("waxing"). Plant crops which produce above ground (e.g., maize, beans, wheat).

Never was the moon said to be observed in determining the time of harvest. Although there may be an idealized moon/harvest relationship similar to the one outlined above for planting (or perhaps the harvest relationship is an inversion of the planting one), the time of harvest is best determined by directly observing the crops and the weather. Planting is a more anxious time, when a number of decisions must be made concerning such meteorological factors as rainfall and temperature. Thus, every available clue, including the moon, is used for deciding when best to plant.

4. Meteorological Lore

Brilliant lines of celestial light, such as rainbows and lightning, unite the three domains of the cosmos (the sky, the earth, and the underworld). Because these phenomena are associated with the atmosphere, and with disturbances of the atmosphere, it is appropriate in the discussion of them to also include shooting stars (small meteors which enter the earth's atmosphere) and the practice of divination by observing the scintillation of stars (scintillation being accounted for in Western meteorology by the irregular fluctuations of density in the atmosphere). For comparative readings on meteorological beliefs, see Boyer (1959), Oesch (1954), Rodríguez (1965), Mariscotti de Görlitz (1973) and Valcárcel (1943).

K'uychi–"Rainbow"

In a number of Quechua dictionaries, the term *k'uychi* is defined not only as "rainbow" but also as the "seven colors" of the rainbow (J. Lira, *Diccionario Kkechuwa Español*, n.d., and R. Aguilar Páez 1970). The notion of seven colors has long been fixed in Western meteorological lore. It is related to the tradition of classical and medieval numerology in which seven is the ideal numerical expression of the basic elements in astrological, alchemical, and religious symbolism. However, it is not at all certain that the Quechua-speakers of today share this notion of the cosmological utility of the number seven nor that the Incas of pre-Columbian times did. In fact, the "seven colors of the rainbow" are a product of the imagination. The colors of a rainbow merge together gradually and one can divide them into any number of categories depending on the purpose for which the division is made (any purpose from religious symbolism to the expression of the frequency of light waves). Thus, it cannot be assumed that seven is necessarily a "logical" Quechua number. For example, Pachacuti

Yamqui (1950:226) uses only four lines in his representation of a rainbow. This drawing may only be a "schematic" of a rainbow, but it suggests that the number seven was not absolutely standardized for the depiction of rainbows in Incaic meteorology and cosmology.[1]

Each informant with whom I discussed the number of colors in a rainbow gave both a different sequence and a different number. One informant said that there are six "classes" of rainbow colors, but he was only able to name five: yellow, red, scarlet, blue, and green. A female informant gave the following list of colors (I have supplied the Quechua terms when given, otherwise the terms were Spanish): blue, yellow (*k'ello*), red (*puka*), rose, dark green (*yanak'omer*), orange (*bella api*), scarlet, sky blue, green (*k'omer*), and the red of the *llaulli* flower (*llaulli puka*). After naming these ten colors, the informant insisted that there are more but she could not remember their names. According to another informant, rainbows have seven colors; however, he did not supply color names. Juvenal Casaverde Rojas (1970:171) was told that rainbows have twelve colors. Obviously, neither the number nor the sequence of rainbow colors is rigidly fixed in Quechua meteorological thought, at least not to the point at which their naming is a simple matter of reciting the proper rhyme or formula.

Informants also give varying descriptions of the shape of rainbows. One informant said that they are in the shape of arcs. Another, describing a vertical circle in the air with his finger to demonstrate, said that rainbows are circles (*muyu*). Rainbows are in fact circles, but a complete circle is almost never visible because much of the rainbow is underground (Graham 1975:75-83 and Nussenzveig 1977). However, one way to see a complete rainbow circle is by looking into a valley from a high mountain, and this may account for the difference between the two informants' descriptions. The one who described the rainbow as an arc lives in the community of Misminay, where the view to the north, west, and south is blocked by mountains, and the view to the east is across a broad flat plain. Thus, rainbows seen from Misminay will commonly be seen as arcs. The second informant lives in Sonqo, in the high puna, and perhaps has had the frequent experience of seeing complete rainbow circles.

Informants were unanimous on two points: rainbows are serpents (*amarus*) which rise out of springs (*pukius*) when it begins to rain (see also Casaverde Rojas 1970:171, J. Núñez del Prado 1970:88, and Vallée 1972:245). Since the Andean rainy season lasts for about six months (November-April), these multicolored, atmospheric serpents are associated with the rainy half of the annual cycle. In this connection it is interesting to note that the dark cloud constellation called Mach'ácuay ("serpent"; *SC#* 41) is said to dominate the night sky during the rainy season. Thus, in the same season that the multicolored rainbow serpents are visible during the daytime, the black serpent is visible at night.

Rainbow serpents are said to have two heads, regardless of whether they are thought of as circles or arcs. The rainbow emerges from a pukiu, arcs through the sky, and buries itself either in the earth or in another pukiu. When a rainbow rises out of one spring and enters another which is at some distance, the two springs are said to be the "same spring." This seems good evidence for the general conception of the rainbow as a complete circle. Two fountains or springs joined by a rainbow can be thought of as the "same" spring in the sense that they are opposing points on a circle; they are the two points which orient the movement of the rainbow downward into the earth and upward into the sky.

After a rainbow stretches across the sky, it does not always remain "disinterested," for rainbows can move willfully along the earth for a variety of reasons, most of which are malevolent. Rainbows move in order to steal (usually from men) and to enter the abdomen of women through the vagina, causing severe stomach pains. Juan Núñez del Prado (1970:88) says that rainbows which arise from subterranean waters cause "intense pains of the stomach and head, vomiting and sickness in general." Both men and women are prohibited from urinating when a rainbow is visible because the rainbow can move across the earth and enter the stomach by way of the urine, again causing severe stomach pains. When a rainbow enters the human body, a direct connection is established between the human body and the subterranean water within the pukiu from which the rainbow emerged. Lira (1946:34) says that the waters from which rainbows arise have small, colorful deposits (*colorcitas*) which can enter the body; if these waters should enter the body, one's urine will be rainbow colored.

As is common in other areas of Quechua symbolism, atmospheric phenomena are conceived of in a sexual dichotomy; rainbows can be either male or female or a male-female combination.[2] One type of rainbow is called *urkuchinantin,* a term which denotes androgynous characteristics. The word is used to refer to a male and female unit or to a female with her male child (cf. Lira, n.d.). Female urkuchinantin rainbows are red, male ones are blue. Red and blue are thought of either as the dominant color of a multicolored rainbow or as the only colors seen in a bow.

All-red rainbows can only be seen moments before and a few minutes after sunset (Minnaert 1954:182-183). Since the sun is thought of as male in Quechua sexual symbolism, red rainbows may be classified as female because they "survive" beyond the setting of the sun and into the night, a period associated with the moon which is female. Urkuchinantin rainbows are especially dangerous to women. They can move across the earth and enter the lower abdomen through the vagina and cause pain or death (cf. Oesch 1954:6 n.1).

Besides androgynous urkuchinantin rainbows, there is another type, which is completely male. These are called *wankar k'uychi* (wankar = "small [war] drum"; k'uychi = "rainbow"). The playing of drums during the rainy season hoeing of maize should be recalled in connection with the association of drums and male rainbows. Also important is the symbolism of drums in Inca times as they relate to meteorological, seasonal, and agricultural phenomena. For example, we have Pachacuti Yamqui's description (1950:227) of the eight large drums which the Inca Lloque Yupanqui ordered to be made, and played with special songs (*ccayo tinmaayma uallina*), at the time of the feast of Capac Raymi (the celebration of the December solstice which falls midway through the rainy season).

According to Casaverde Rojas (1970:171), wankar k'uychi has "a large head which is fastened to the rainbow like a tail, large ears, dark cloudy eyes, eyebrows, eyelashes, and a beard and white hair." These multicolored male rainbows are usually seen high in the sky, their lower parts being obscured by heavy clouds. Along the high, visible part of wankar k'uychi a black cat (*k'owa*) can be seen moving stealthily up the side of the bow. This black cat is a repugnant creature in Quechua thought and is considered to be a demon or evil spirit (Earls and Silverblatt 1978:322). The k'owa can be either male or female. Mishkin (1940:237) found in Kauri that the Ccoa (= K'owa) is referred to as the "cat of the apu"; it is the harbinger of hail and lightning. In Cailloma, the k'owa, or *chinchaya*, is also associated with thunder and lightning (Christopher Wallis, personal communication, 1977).

Wankar k'uychi is especially despised because he is a thief. If someone is wearing a blue scarf, for instance, wankar k'uychi can move by, or through, the person and steal the scarf. It is not clear to me whether the theft is committed by the black cat or by the bands of the rainbow which match the color of the stolen article of clothing. The theft of clothing is described as a very rapid, almost magical flight into the sky. One informant who discussed the thievery of wankar k'uychi consistently referred to red and blue articles of clothing to describe how the thefts take place. Thus, red and blue appear to be the primary colors in the classification and conceptualization of rainbows.

Chimpu—"Haloes"

While the term *k'uychi* specifically refers to rainbows, it can also be used generically for other rainbow-like phenomena. One informant identified all atmospheric bows (rainbows, haloes, and coronas) as k'uychi. However, haloes are more frequently referred to as *chimpu* ("halo, aureole" or "mark"). *Chimpu* is also the term used in the area of Ayacucho for the colorful pieces of yarn looped through the pierced ears of llamas and then tied (R. T. Zuidema, personal communication, 1978). Solar haloes are called *intita chimpushan k'uychi* ("the rainbow marking the sun"). The informants who gave this identification for solar haloes emphasized the fact that haloes are "fixed marks" around the sun. Lunar haloes in Misminay are called *quillata chimpushan k'uychi* ("the rainbow marking the moon").

The notion of a rainbow "marking" the sun or moon is related to the belief that solar and lunar haloes presage rain.[3] The prediction of rain by haloes is done by judging the relative size of the halo: small ones precede rain by a number of days, larger ones by only a day or two. In this context, it is interesting to note that the word *chimpu* is also used to designate a line or string with which a bag is marked to indicate the quantity of goods inside the bag (Lira, n.d.). Thus, the larger the quantity of goods in a bag, the larger must be the string to mark it; similarly, larger haloes "mark" or signify larger amounts of moisture in the air. The conceptual correlation between haloes and water is evident also in the community of Cailloma, where solar haloes are referred to as *inti qhuchapi* ("sun in the lake"; C. Wallis, personal communication, 1977).

In relation to the prediction of rain, solar haloes are said to have only three colors, but there are three colors for solar haloes of the rainy season and another three for those of the dry season. My informant was able to name only two of the dry-season colors: light blue and yellow. Those of the rainy season are purple, yellow, and "medium-white" (*medio blanco*). Perhaps these different colors, because they relate to periods of raininess and dryness, are thought to contain or represent varying quantities of "color-moisture." Rainy-season and dry-season haloes were not described to me in terms of a sexual dichotomy, but since rainbows and lightning have a male/female classification, the same is probably true for haloes: dry season = male haloes; rainy season = female haloes.

Illapa, Rayo, and *Relámpago*–"Lightning"

Illapa is a generic term referring to all forms of lightning. The word is known in the community of Misminay but is used less frequently than two Spanish terms: *rayo* ("lightning" or "thunderbolt") and *relámpago* ("lightning" or "flash"). These terms represent two different types of lightning: rayo refers to female lightning, Relámpago to male lightning (although I have also heard rayo used both for male and female lightning in much the same way that urkuchinantin is used for androgynous rainbows).

Female lightning is vertical lightning which strikes very quickly, almost without noise, straight down into the earth. Rayo is a thief and harmful to women. As for the association with women, rayo is said to kill or carry away women who are alone in the mountains pasturing sheep or cattle. It apparently has the power neither to kill nor to carry away men. During my residence in Misminay, a woman was said to have been struck by rayo while pasturing cattle. The day she was struck, a female informant told me that the woman had been badly injured by lightning; some eight months later, a male informant repeated the story but said that the woman had been carried away by rayo. He insisted that I not be concerned because she had been, he said, a notoriously bad woman. The general impression given is that women who pasture livestock in the mountains alone place themselves in a very precarious position and are inviting some form of catastrophe.

The idea that rayo is a thief is related to the belief that, when she strikes the earth and enters it a short distance, she enters at those places where someone has hidden silver or coins. Rayo steals the hidden treasure and takes it up to the sky. Shooting stars (*ch'aska plata*–"silver hair" [star]) are another source of subterranean silver stolen by rayo. Female lightning is also considered to be a "demon" and a "cat" (*saqro*); it is therefore probably related to the black cat (k'owa) that accompanies the thieving wankar k'uychi.

Male lightning (relampago) is the form of lightning associated with loud rumblings. It strikes with a great clap of thunder but does not come to earth. It either stays in the clouds in the form of horizontal flashes of light, or it strikes midway between the sky and the earth at a diagonal angle.

Sullaje

Another luminous phenomenon associated with rain is *sullaje*, a word derived from *sulla* ("shower" or "spray"). Sullaje is a flash, or a series of flashes of light which move horizontally along a distant range of mountains during a rainstorm. Sullaje is always *k'ello* ("yellow"), the color of fire and of the sun. The informant who described this phenomenon insisted that it is not the same thing as lightning.

Ch'aska Plata or *Boleadora*–"Shooting Stars"

The term *ch'aska plata* ("silver star") is applied to shooting stars, although they are also referred to by the Spanish term *boleadora* (a lasso with three balls at the end). During Inca times, the boleadoras were made of bronze and silver with incrustations of silver and copper (Larrea 1960:97). The "silver" of shooting stars is one origin of the silver taken out of the ground by female lightning when it penetrates the earth.

A form of divination is carried out by observing shooting stars. If a priest (paqo, or altomisayuq) has been asked to divine the whereabouts of a stolen article, for instance, he watches the sky and divines the location of the article, or the direction to the house of the thief, on the basis of the point of origin and trajectory of a shooting star. Some priests are able to divine for such things as death, illness, and theft by interpreting the final burst of light given off by a shooting star.

Stellar Divination

Several forms of divination are based upon the observation of the scintillation of starlight. One informant gave the following example. If someone capable of divination is traveling at night, he determines which star stands in the zenith of a distant house. By interpreting certain (unspecified) characteristics of the scintillation of the star in question, he is able to divine for such things as death or theft for the occupants of the house. This form of stellar divination requires the ability to interpret the specific meaning of various types of starlight, its color and degree of scintillation, for instance. Another form of stellar divination and prognostication is carried out by observing the relative brightness of scintillation of the Pleiades. The several methods of divination by observing the Pleiades are discussed in chapter 6.

Relating Meteorology to Cosmology

The meteorological phenomena discussed in this chapter share several important features. Principal among these are the relationship between light, color, and water and the fact that these phenomena are all regarded primarily as manifestations of chthonic or celestial forces which, by their emergence from the earth or the sky, establish an interconnection among the various levels of the universe. Rainbows are especially important in this regard because they are thought of as giant serpents which rise up out of subterranean springs, pass through the sky, and bury one of their two heads in a distant spring. We will find later that mountain springs are also thought to be the places of origin of animals on the earth as well as the principal places of entry into the interior of Pachamama (Earth Mother). Rainbows are therefore the manifestation, in a reptilian form, of the forces of procreation and fecundity (*pachatira*) which lie within the earth. In short, we may conclude that all meteorological phenomena are reflections of the presence and action of powerful natural forces, primarily those forces associated with *water*, the source of fertility.

Since the Milky Way is thought of as a river which transports water from the cosmic sea to the celestial sphere, a relationship exists between the rainbow/serpent, which arches through the sky during the daytime, and the Milky Way, which arches through the night sky. The relationship of rainbows, the Milky Way, and a dark streak in the Milky Way which is also referred to as a celestial serpent are more closely examined in chapter 9 but, before leaving the subject of rainbows, it should be pointed out that there is an orientational relationship between the arcs of rainbows and the Milky Way with respect to the changing position of the sun as it moves from solstice to solstice. The orientational data outlined below incorporate rainbows into the cosmological structures and orientational principles discussed earlier.

In figure 22 we saw that there is a coincidence in the orientation of the points of sunrise of the two solstices and the orientations of the two axes of the Milky Way during the early evenings at the times of the solstices. However, on the evening of the June solstice when the sun *sets* in the northwest, the Milky Way will begin to appear across the sky in a line from the northeast to the southwest (i.e., it will form an arc *opposite* the sun); at the time of the December solstice, the sun will set in the southwest and the Milky Way will be seen in a line running from the northwest to the southeast. Thus, on the evenings of the two solstices, the Milky Way will be seen in opposite directions from the sun. This, in fact, is exactly the same positional relationship that exists between rainbows and the sun; they are always seen as arcs (or circles) stretching across the sky directly opposite the sun (see Minnaert 1954). Therefore, rainbows and the Milky Way may be equated not only because they are continuous arcs passing through the sky, but also because there is a consistently observable relationship between the sun and the celestial arc.

5. The Stars and Constellations

The list in table 7 contains the stellar and planetary data I collected in the field from 1975 to 1977. It could be expanded by including stellar data reported in the ethnohistoric and ethnographic literature. However, although I refer to comparative ethnoastronomical data throughout the text and list the identifications made by other Cuzco-area ethnographers in the Appendix, for the purposes of this study it is necessary to place a minimal control on the data to be analyzed. It is often impossible to know the precise conditions under which an ethnographer has collected ethnoastronomical data—for instance, the exact time when stellar identifications were made. This information could be important because often stars are not given fixed names; instead, they are given a number of different names depending on the different celestial *positions* they occupy (see chap. 8). Thus, a specific name may be attached to a stellar *function* or celestial *location.* Examples are the terms "morning star" and "evening star," which are applied to *any* planet or bright star occupying a certain celestial position at a certain time; the two terms do not necessarily refer to Venus.[1]

In addition, I have found that even when an informant does point to a star or group of stars, it is often impossible to know for certain that you are indeed both looking at the same thing. This is especially critical in Quechua astronomy because an informant may point to a black spot (i.e., a dark cloud constellation), but the anthropologist, being unfamiliar with "negative constellations," may assume that the informant is identifying the stars which are near to, or encircling, the black spot. This is further complicated in the Andes by the fact that one encounters a general disinclination to point to *anything* considered sacred; among other things, this includes mountains, stars, and rainbows. The ways of getting around the problem of pointing directly at something are ingenious and often frustrating. A sacred mountain, for instance, may be pointed at by holding your hands behind your back and gesturing with one elbow; stars were sometimes

Table 7. A Catalogue of Quechua Stars and Constellations

	Name	Translation	Provenience	Identification	Remarks
			Single Stars and Planets		
+*	1. *Alto piña ch'aska*	"star which is already high" [i.e., in the zenith]	Songo Yucay	Various 1st and 2nd magnitude stars and a planet	A star or planet in the "zenith" at midnight; chap. 8.
+	2. *Boleadora*	(Sp.) a lasso with three balls at the end; also a "thief"	Misminay	Shooting stars	Chap. 4.
+	3. *Ch'aska plata*	"silver star"	Misminay	Shooting stars	Chap. 4.
+	4. *Ch'issin ch'aska*	"evening star"	Misminay Sonqo Yucay Quispihuara	Various 1st and 2nd magnitude stars and a planet	The evening star; chap. 8.
?	5. *Coscotoca ch'aska*	"star of the two equal parts"	Sonqo	Various 1st and 2nd magnitude stars and a planet	A star or planet in the "zenith" at midnight; chap. 8.
?	6. *Hatun Coyllur*	"large star"	Sonqo	Sirius	This may be a generic term, equivalent to our "1st magnitude."

Table 7 (continued)

	Name	Translation	Provenience	Identification	Remarks
			Single Stars and Planets (continued)		
+	7. *Illarimi ch'aska*	"dawn star"	Songo Yucay Quispihuara	Various 1st and 2nd magnitude stars and a planet	Morning star; chap. 8.
+	8. *Locero*	(Sp.) "morning star"; Venus	Songo Yucay Lucre	Various 1st and 2nd magnitude stars and a planet	In Yucay, one informant talked of *pachaypaqa locero* ("dawn morning star"); an informant in Lucre called both the morning and evening stars "*locero*."
+	9. *Pachapacariq ch'aska*	"dawn of the earth/time star," "morning star"	Misminay Songo Yucay	Various 1st and 2nd magnitude stars and a planet	One Songo informant re-referred to *pachaillarimi ch'aska* ("dawn of the earth star"); chap. 8.
?	10. *Torito*	(Sp.) "little bull"	Misminay	Various 1st and 2nd magnitude stars and a planet	A star or planet in the "zenith" at midnight; *torito* is also the name of the "rhinoceros beetle" *(Dynastinae)*; chap. 8.
			Star-to-Star Constellations		
+	11. *Amaru contor*	the serpent changing to the condor	Chumbivilcas	Scorpio	Fig. 32.

Table 7 (continued)

	Name	Translation	Provenience	Identification	Remarks
				Star-to-Star Constellations (continued)	
−	12. *Arado*	(Sp.) "plough"	Lucre	Scorpio (?)	
+	13. *Boca del Sapo*	(Sp.) "mouth of the toad"	Misminay	The Hyades	The informant said that it is shaped like [drawn curve] and that it rises about 8:00 p.m. in mid-September; chap. 6.
−	14. *Cabañuelas*	(Sp.) "small hut, cottage"	Cuzco	The Pleiades (?)	A group of small stars which rise on August 1 and are used for crop predictions; chap. 6.
+	15. *Calvario Cruz*	(Sp.) "the cross of Calvary"	Misminay Sonqo	(a) head of Scorpio (b) Betelgeuse, Rigel, Sirius, and Procyon (c) Procyon, Castor, Belt of Orion, ß Tauri (d) Belt of Orion, τ Orionis, and Rigel	Chap. 7.
+	16. *Collca*	"storehouse"	Misminay Quispihuara Sonqo Yucay	(a) the Pleiades (b) the tail of Scorpio (c) the Hyades	Observed in weather and crop predictions; chap. 6 and fig. 32.

Table 7 (continued)

Star-to-Star Constellations (continued)

	Name	*Translation*	*Provenience*	*Identification*	*Remarks*
?	17. *Contor*	"condor"	Misminay	(a) the head of Scorpio (b) head = δ, ε, n Canis Majoris wing = α, ß, γ Pyxidis wing = α Monocerotis tail = α Hydrae	Chap. 7 and no. 36.
+	18. *Chakána*	"bridge, stair"	Misminay Songo	(a) Belt of Orion (b) Chacanuay = δ, ε, η Canis Majoris	Chap. 7 and fig. 32.
?	19. *Choque-chinchay*	"golden cat"	Songo	Tail of Scorpio (or dark spot inside tail?)	Chap. 6.
+	20. *Hatun Cruz*	"great, or large cross"	Misminay Songo	(a) Betelgeuse, Rigel, Sirius, Procyon (b) Procyon, Castor, η and μ Geminorum	In Misminay, *Hatun Cruz* is the cross of the northern *suyu* ("quarter"); chap. 7 and fig. 32.
+	21. *Huchuy Cruz*	"small cross"	Songo	The Southern Cross (*Crux*)	In Yucay, the Southern Cross is called *Calvario;* chap. 7 and fig. 32.

Table 7 (continued)

Star-to-Star Constellations (continued)

	Name	*Translation*	*Provenience*	*Identification*	*Remarks*
?	22. *Khaswa Coyllur*	"circle dance star(s)"	Sonqo	(a) $\gamma, \delta, \kappa, \lambda$ Vela (b) Corona Borealis (?)	The informant made a circular, spiraling motion when identifying *khaswa coyllur* on a star map.
?	23. *Llama Cancha*	"Llama corral"	Misminay	?	Informant described *Llama cancha* as a group of 56 stars.
+	24. *Llamacñawin*	"eyes of the llama"	Misminay Yucay Sonqo Lucre Quispihuara	α and ß Centauri	The "eyes" of the Dark Cloud constellation of the *Llama;* Fig. 32 and no. 40.
+	25. *Linun Cruz*	*Linun* = Lat. *lignum;* "wooden cross" or "the crucifixion cross"	Misminay	The head of Scorpio	Chap. 7 and fig. 32.
–	26. *Mamana micuc*	"(the one who) eats his mother"	Misminay	$\xi, \pi, \sigma, \nu, \tau$ Puppis	A zig-zag line of stars near the mouth of the Dark Cloud "Serpent" (Mach'ácuay); informant's drawing:

Table 7 (continued)

Star-to-Star Constellations (continued)

	Name	*Translation*	*Provenience*	*Identification*	*Remarks*
?	27. *Ñawin Cristo*	"the eyes of Christ"	Misminay	α and ß Centauri (?)	One informant mentioned *ñawin Cristo* while discussing the *Llama* (*llamacñawin* = "eyes of the Llama")
+	28. *Papa Dios Cruz*	(Sp.) "the cross of God the father"	Misminay	Head of Scorpio	*Papa Dios Cruz* = *Calvario Cruz* (no. 15); chap. 7 and fig. 32.
?	29. *Passon Cruz*	(Sp.) "the cross of the Passion"	Lucre	?	Informant's drawing:  chap. 7.
?	30. *Pisqa Collca*	"the five storehouses"	Sonqo	Located somewhere from the belt of Orion to the Hyades	See no. 32; chap. 6 and fig. 32.
+	31. *Pisqa Coyllur*	"the five star(s)"	Misminay	The Hyades	One informant said that *Pisqa collca* (no. 30) = *Pisqa Coyllur.*
+	32. *Qutu*	"pile"	Sonqo	The Pleiades	The Pleiades in Sonqo are also called *collca qutu* ("storehouse pile"); chap. 6 and fig. 32.

Table 7 (continued)

	Name	Translation	Provenience	Identification	Remarks
				Star-to-Star Constellations (continued)	
?	33. *Sonaja*	Sp.) "tambourine, rattle"	Misminay	(?) Corona Borealis	*Sonaja* was said to set in the northwest about where the June solstice sun sets: Solstice set = 296°; Corona Bor. set =298°
+	34. *Yutucruz*	"tinamou cross"	Chumbivilcas	The Southern Cross	*Yutu* = the Coalsack (no. 47).
				Dark Cloud Constellations	
+	35. *Atoq*	"fox"	Misminay	Dark spot between the tail of Scorpio and Sagittarius	Chaps. 2 and 9 and fig. 33.
?	36. *Contor*	"condor"	Misminay	Dark spot *perhaps* in the area of Scorpio	Chap. 9 and no. 17.
+	37. *Hanp'átu*	"toad"	Misminay Sonqo Yucay Lucre Quispihuara	(a) small dark spot SW of the Southern Cross (b) the Coalsack, a dark spot SE of the Southern Cross (c) dark spot outlined by the tail of Scorpio	Cf. no. 13 and 48. Fig. 33 and chap. 9.

Table 7 (continued)

Dark Cloud Constellations (continued)

	Name	*Translation*	*Provenience*	*Identification*	*Remarks*
+	38. *Hatun llamaytoq*	"the great, large llama"	Misminay	Dark streak between the Coal-sack and ε Scorpii	*Hatun llamaytoq* = *Llama* (cf. no. 40). Chap. 9
–	39. *K'olli*	?	Sonqo	Unidentified small dark spot	
+	40. *Llama*	llama	Misminay Sonqo Yucay Lucre Quispihuara	Large dark streak between the Coalsack and ε Scorpii	Chaps. 3, 4, and 9, and fig. 33
+	41. *Mach'ácuay*	"serpent"	Misminay Sonqo Yucay	(a) large *S*-shaped dark streak between Adhara and the Southern Cross (b) dark streak from α Centauri south and curving up to θ Scorpii	Cf. no. 42b, chap. 9 and fig. 33.
+	42. *Mayu*	"river"	Misminay Sonqo Lucre Yucay Quispihuara	(a) the Milky Way (b) one informant said that the *Mayu* is the large *S*-shaped black streak between Adhara and the Southern Cross (cf. no. 41a)	Chap. 2 and 9 and fig. 33.

Table 7 (continued)

Dark Cloud Constellations (continued)

	Name	Translation	Provenience	Identification	Remarks
?	43. *Ombligo de la llama*	(Sp.) "umbilicus of the llama"	Chumbivilcas	The narrow, curving dark streak from α Centauri south and curving up to θ Scorpii	Cf. no. 41b and no. 45 located beneath the *Llama*); chap. 9 and fig. 33.
–	44. *Sullu ullucu*	"aborted *ullucu*"	Sonqo	?	*Ullucu* = *Ullucus tuberosus*, a high-altitude tuber; informant associated it with Mach'ácuay, no. 41; see chap. 9.
+	45. *Uñallamacha*	"suckling baby llama"	Misminay	The narrow, curving dark streak from α Centauri south and curving up to θ Scorpii	Cf. no. 41b and no. 43 (located beneath the *Llama*); chap. 9 and fig. 33.
+	46. *Urpi*	(Sp.) "dove"	Yucay	Coalsack	*Urpi* = *yutu* ("tinamou," cf. no. 47); chap. 9.
+	47. *Yutu*	"tinamou"	Misminay Sonqo Yucay Lucre Quispihuara	(a) Coalsack (SE of Southern Cross) (b) dark spot outlined by the tail of Scorpio (c) dark spot near Scutum	Chaps. 5 and 9 and no. 46 one informant referred to the Coalsack as *yutucha* ("little tinamou"), fig. 33.

Table 7 (continued)

Light Cloud Constellations

	Name	*Translation*	*Provenience*	*Identification*	*Remarks*
?	48. *Hanp'átu*	"toad"	Sonqo	Bright nebulosity in the "curve" of Scorpio (i.e., between the head and the tail of Scorpio)	The informant was uncertain as to whether Hanp'átu is this "light cloud" or if it is the dark cloud of 37c.
?	49. *Mama Rosario*	"Mother Rosary"	Misminay	Unidentified bright nebulosity in the Milky Way	
+	50. *Posuqu*	"foam"	Misminay	Bright stellar clouds in the southern Milky Way	Represents the "foam" of the collision of the two celestial rivers; chap. 2.

*Symbols in this column indicate the degree to which confidence in the reliability of the data is justified: + = confident; ? = questionable; — = not confident.

identified for me in the sky by a wave of the hand or the wriggle of a couple of fingers. This was occasionally carried to the extreme of looking at the stars in the sky but actually pointing at them on a star map.

On one occasion, an old man who was very interested in astronomy agreed to discuss the stars with me, but only if I remained seated on the floor inside his hut during the interview. He began the interview, which took place at about midnight on a clear, moonlit night, by first instructing me to sit on the floor in the center of his hut; he and his son then went outside to look at the sky. As the interview proceeded, the two of them would occasionally go outside to count the stars in a constellation or to refresh their memories about where certain constellations were located at the moment. Throughout the interview, he refused to allow me to accompany him outside. It should be mentioned that I was not well acquainted with the informant at the time of this interview, and his disinclination to go outside with me may have been based more on a fear or uncertainty about being outside at night with *me* rather than an unwillingness to point at the stars.

Because of these and similar difficulties in collecting astronomical data, I have chosen to concentrate on the material with which I am most familiar and for which I am able to make occasional judgments concerning the reliability of the information. This brings up another point. On countless occasions during one's fieldwork it is necessary to decide how extensively a particular datum must be corroborated, either by discussing it with different informants or by asking the same informant the same question at different times. Corroborating every bit of data is obviously desirable but not always possible. This proved to be especially true in collecting ethnoastronomical material. Some informants became progressively more inclined to discuss astronomical topics with me, often to the point of volunteering interviews, but a considerable amount of data were collected under less favorable conditions (e.g., discussing the stars during the daytime or finding oneself in the position, frequent in the Andes, of discussing the universe while in an extreme state of inebriation). Thus, for each item in the catalogue I have indicated in the column on the extreme left how confident I am in the general reliability of the data (especially the identification). The degree of confidence is based on my understanding of the stellar identifications, the number of corroborations for each identification, and my general impression of the reliability and sophistication of the informant(s).

The Star and Constellation Catalogue is divided into four categories. These are (a) single stars and planets, (b) Star-to-Star Constellations, (c) Dark Cloud Constellations, and (d) Light Cloud Constellations.

Single Stars and Planets

Several informants stated that all the stars, regardless of their size or color, can be referred to either as *ch'aska* ("shaggy hair") or *coyllur* ("star"). However, the term *ch'aska* seems to be used primarily for a planet, or for a first-magnitude star "replacing" a planet, which appears as the morning, evening, or "zenith" midnight star (see chap. 9). *Coyllur* is used primarily for stars of lesser magnitude, or for first-magnitude stars when they are *not* equated with planets. In the community of Cailloma (Department of Arequipa), Christopher Wallis found that *coyllur* is the general term used for bright stars, whereas *ch'aska* refers to the brightest ("abundant") stars (personal communication, 1977). Thus, the terms *ch'aska* and *coyllur* are used either to distinguish *stars* of different relative magnitudes, or, in another context, to distinguish bright stars and planets which occupy an important *position*, such as morning star or evening star, from the rest of the stars.

Stars are primarily classified as masculine. Related to this characteristic is the fact that in the historical/cosmological order of the creation of the universe, the stars appeared before either the sun or the moon. This is especially true in the case of *papa pachapacariq ch'aska* ("father morning star"), which is said to have been the first bright object to appear in the dark primeval sky at the beginning of time. The single stars and planets, which together I refer to as the "twilight stars," are discussed in detail in chapter 9.

The Spanish and Quechua terms for "father" are often used in references to celestial phenomena. However, most commonly one finds the Spanish *papa* used in relation to the stars and the Quechua *taytay* (*taytacha*) used in relation to the sun. I did not hear the other principal Quechua term for "father," *yaya*, used in an astronomical context.

Star-to-Star Constellations

The phrase "star-to-star" is one I have coined; it was not used by an informant. In fact, there does not appear to be a common generic term for this type of constellation. The star-to-star constellations (fig. 32) are similar to Western European constellations in that they trace a familiar shape in the sky by conceptually linking together neighboring bright stars. Most of the constellations of this type are found along or near the central path of the Milky Way, but they are especially prominent in the region of Taurus and Orion, where the Milky Way has its greatest width and lowest surface brightness. Because of the greater width of the Milky Way and the lesser density of stars in this part of the sky, individual bright stars and small clusters of lesser magnitude stars (e.g., the Pleiades and the Hyades) are more easily distinguished. In general, star-to-star constellations are either geometrical (the Large and Small Crosses) or they represent inanimate, usually architectural objects (the Bridge and the Storehouse).

Star-to-star constellations are primarily classified as masculine; they are not

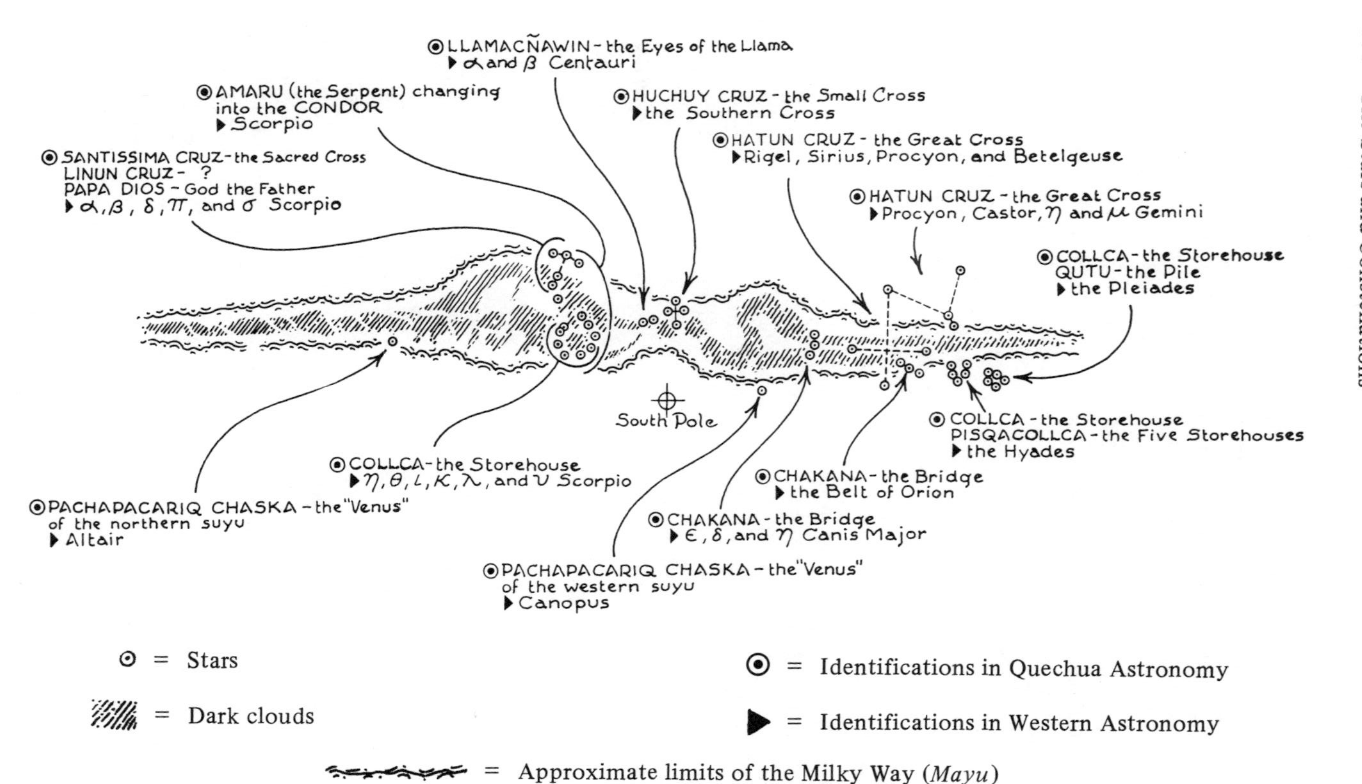

Fig. 32. The Star-to-Star Constellations

uncommonly referred to as *papa* . . . ("father . . ."). However, a note of caution must be entered here. It seems to be a common feature of Quechua classifications that objects or categories do not have absolute symbolic values. That is, one must always consider the context within which classifications are made. Thus, although star-to-star constellations as a *category* are considered masculine, a specific constellation, or one star of a constellation, may be masculine, feminine, or androgynous. The large group of star-to-star constellations representing crosses is discussed in chapter 7; those representing storehouses (e.g., the Pleiades and the tail of Scorpio) are discussed in chapter 6.

Dark Cloud Constellations

In addition to single stars, planets, and star-to-star constellations, another large group of celestial objects is confined entirely within the bounds of the Milky Way (the Mayu). Elsewhere (Urton 1978a and 1978b), I have referred to these celestial objects as "*black* constellations," but a more appropriate name, both because it is the indigenous term and because it better describes these objects, is "dark cloud" (*yana phuyu*). The word *yana* is sometimes translated as "black." However, in the Quechua conception of light or color classifications, yana is thought of as "dark" (or "obscure") in opposition to "light," rather than as black opposed to white.

Dark cloud constellations (fig. 33) are located in that portion of the Milky Way where one sees the densest clustering of stars and the greatest surface brightness, and where the fixed clouds of interstellar dust (the dark cloud constellations) which cut through the Milky Way therefore appear in sharper contrast. From the earth, these dark spots appear to be huge shadows or silhouettes pasted against the bright Milky Way. In contrast to the star-to-star constellations, which I have characterized as inanimate, geometrical, or architectural figures, the dark cloud constellations are either animals or plants, usually the former. Dark cloud constellations have been described elsewhere in South America (Lévi-Strauss 1973:134-135 and 1978:110-111; Nimuendajú 1948:265; Reichel-Dolmatoff 1975:115; Tastevin 1925:182, 191; and Weiss 1972:160), among Australian aboriginal populations (Maegraith 1932 and Mountford 1978:59), in Africa (Lagercrantz 1952), and in Java (Pannekoek 1929:51-55 and Stein Callenfels 1931).

Symbolically, dark cloud constellations represent a transitional, intermediate category of celestial phenomena. That is, they are androgynous or asexual, and even though they are located in the sky, they are classified as *pachatierra* (or *pachatira*), a word which combines the Quechua and the Spanish terms for "earth." Chapter 9 discusses the symbolic characteristics of these silhouettes in the Milky Way as well as the correlation of the astronomical periodicities of the dark-cloud animal constellations and the biological cycles of their animal counter-

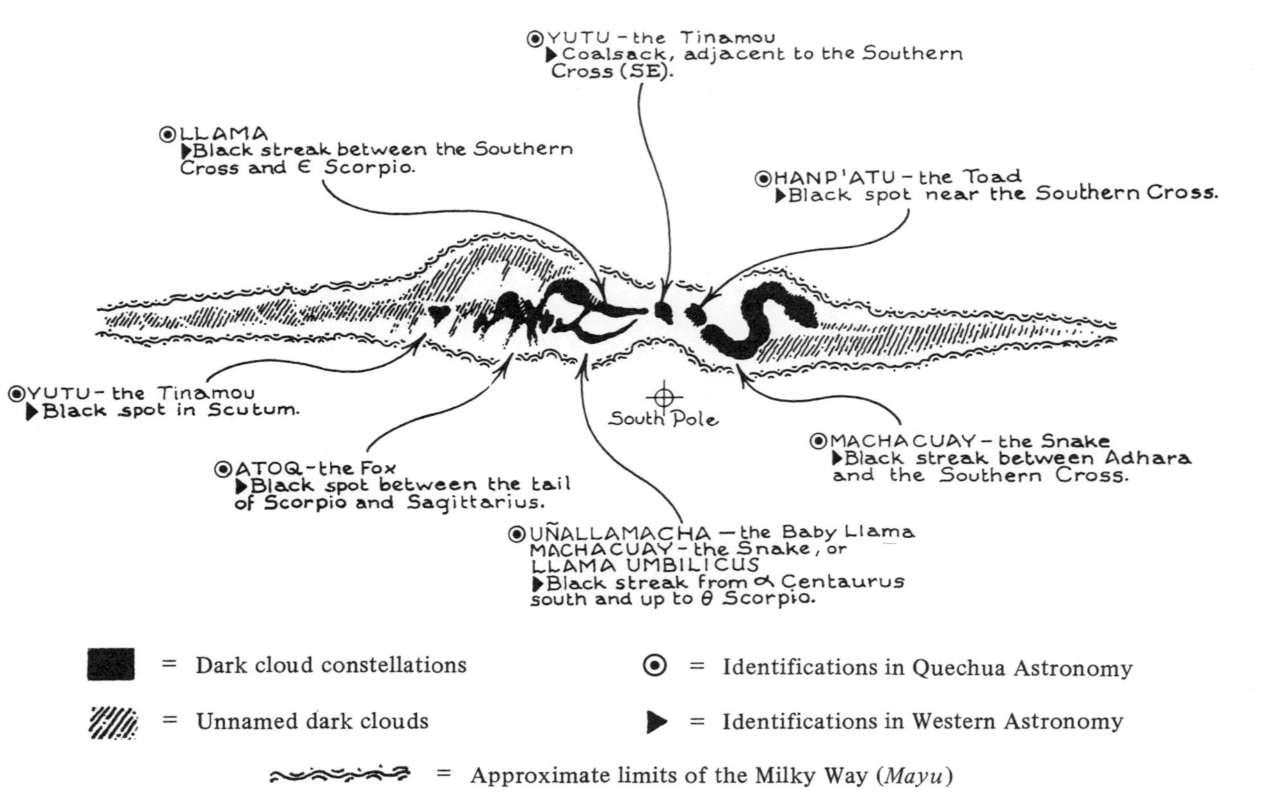

Fig. 33. The Dark Cloud Constellations

parts on the earth.

Light Cloud Constellations

On three separate occasions, informants gave names to various bright nebulae of the southern Milky Way. Since I was not told a generic name for these bright stellar clouds, I have coined the phrase "light cloud," partly to contrast them to the dark cloud constellations. Because I am not certain that this classification would even be intelligible to a runa, I have little to say about these constellations here, the one exception being the posuqu ("foam") of the two colliding celestial rivers, already discussed in chapter 2. I would hypothesize, however, that light cloud constellations are primarily classified as female (e.g., Mama Rosario, *SC* # 49).

We find an interesting suggestion for the possible relationship between dark cloud and light cloud constellations in a myth from the area of Ayacucho. Alejandro Ortiz Rescaniere (1973:90) recounts a myth in which cups of black and white water ascend to the sky from Lake Amaru Cocha: "Of the figures which were elevated from the lake, the white ones represent good and the black ones represent evil." Thus, dark cloud and light cloud constellations may have a common terrestrial origin, the former representing patches of earth (or cups of black water) in the sky, the latter representing cups of white water fixed along the course of the celestial River.

6. Collca: The Celestial Storehouse

A number of informants located the constellation Collca ("storehouse") for me both in the night sky and on star charts. These identifications will be used to establish a range of possible definitions for the term *collca.*

The Pleiades. In most communities, the term *collca* refers to the Pleiades, a small cluster of some six or seven stars in the constellation called Taurus in Western astronomy (see *SC#* 16). Most astronomy textbooks maintain that the naked eye can distinguish only six or seven stars in the Pleiades. However, one informant in Misminay insisted that ten stars are visible in Collca; she identified Collca as the Pleiades. A young man in Lucre said that he did not know how many stars there are in the cluster, but his drawing of Collca contains sixteen stars (although this is possibly only a glyph for "a bunch of stars"). I consider the Pleiades to be the primary referent of the term *collca.*

In Sonqo, the Pleiades are referred to as *qutu* ("pile"; *SC#* 32) and less frequently as *collca qutu* ("storehouse pile"; *SC#* 32). In his dictionary of modern Quechua, Padre Jorge Lira (n.d.) says that *khutu* is an adjective meaning "still, motionless, frozen, and cold." It is possible that this is another translation for the use of the term *qutu* in Sonqo and that it refers to the fact that the Pleiades rise heliacally during the winter (i.e., in June). In the astronomy of the contemporary Aymara-speakers of Bolivia, the Pleiades are also called *qutu* ("handful" or "group"); they are said to represent a group of virgins (Tejeíro 1955:73). Thus, the Sonqueño term for the Pleiades may be derived from Aymara.

In the general Cuzco area, the Pleiades are also referred to by the Spanish phrase *las siete cabrillas* ("the seven goats").

The Tail of Scorpio. In Sonqo, Collca is also identified as the five stars at the tail of Scorpio (θ, ι, κ, λ, and υ Scorpii; *SC#* 32). An informant in Yucay said that Collca has the shape of a cup or bowl. Although the Pleiades can be seen as a tiny dipper with a handle, I am confident that the informant was describing not the handled dipper of the Pleiades but the handleless, well-outlined cup of the tail of Scorpio.

One very reliable informant in Sonqo said that the tail of Scorpio is called both *collca* and *choque chinchay* ("golden cat"). Choque chinchay is also mentioned in the chronicles as an Incaic astronomical term (Pachacuti Yamqui [1950: 226] refers to the evening star as choque chinchay).

The Hyades. One informant identified Collca as the *V*-shaped group of five stars in Taurus called the Hyades (α, γ, δ, ϵ and θ Tauri). However, the Hyades are more commonly referred to as *pisqa coyllur* ("five stars") or *pisqa collca* ("five storehouses"; *SC* # 30, 31). The Hyades are located very near to the Pleiades (they are separated only by some 10°), but as a referent of Collca they are secondary and are subsumed within the far greater role of Collca which is played by their neighbor, the Pleiades.

On the basis of these data, we can conclude that *collca* identifies two basic sets of stars: the Pleiades and the five stars at the tail of Scorpio. Before going further it should be pointed out that a relation between these two sets of stars is not entirely arbitrary. The two "storehouses" are in close opposition to each other in the sky. The Pleiades are located at right ascension (rh) 3 hours 45 minutes, while the center of the tail of Scorpio is at rh 17 hours 15 minutes. Thus, the two collcas are separated by 10 hours 30 minutes in one direction, and 13 hours 30 minutes in the other. This means that the heliacal rise of one Collca will occur within a half month of the heliacal setting of the other.[1] The observational effect of this is that while one Collca is in the sky, the other will be down, and every twelve hours or so one Collca will replace the other in the night sky.[2]

The Opposition of the Two Collcas

It is meaningless, of course, to search a celestial globe for bright stars or groups of stars in opposition. The possible combinations are almost infinite. It is not even so remarkable to find the same word used to refer to one or more of these opposing groups of stars. Named oppositions do not become interesting until it can be shown that the equations go beyond terminology; that the two groups of stars receive the same name, and are thereby formally related, because their opposition expresses something basic about the conceptualization and organization of time and space. The following data would seem to confirm the cosmological and calendrical significance of the Collca opposition in Misminay.

During the course of a long interview on the stars, an informant from Misminay said repeatedly that Collca rises at the city of Urubamba and sets at Apu

K'otin. Figure 34 is a map showing the community of Misminay in the center and the orientations to Urubamba, Apu K'otin, and the rise and set points of the Pleiades and of the tail of Scorpio. There is only about a 2° discrepancy between the orientation from Misminay to the rising point of the Pleiades and the place where the informant said it rises (Urubamba). This is well within reason for an astronomical system which has not constructed cairns or other artificial devices for making precise astronomical observations. However, the setting point of the tail of Scorpio and the highest peak of Apu K'otin are separated by about 20°. Even if we consider the base rather than the main peak of Apu K'otin, there is still a difference of around 15°.

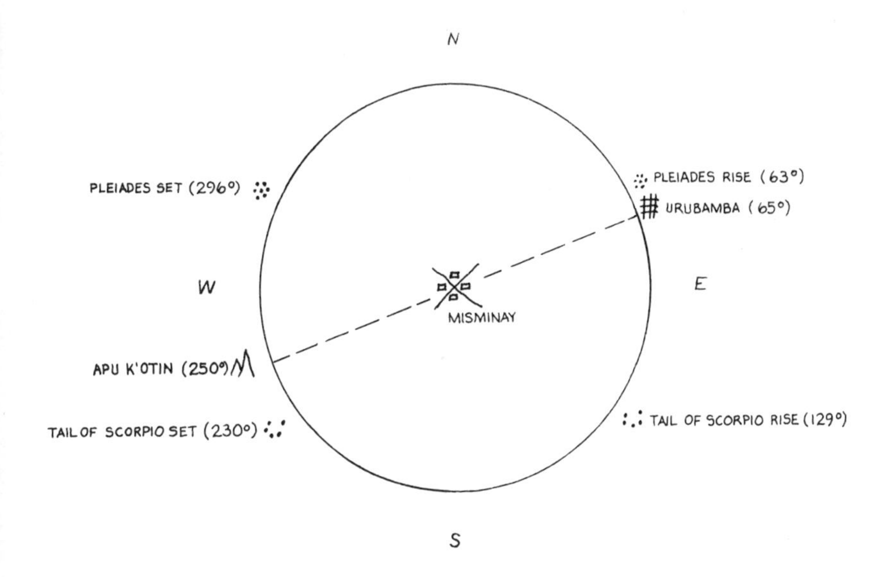

Fig. 34. The Collca Axis

The best explanation for this "error" is that in Misminay, and in every other community in which I have collected astronomical data, it is common to relate the rise and set points of celestial bodies to the nearest important landmark. In the case of the setting point of the tail of Scorpio, the nearest landmark is Apu K'otin; it is therefore "logical" to relate the two. For those familiar with the rather precise degree-of-error allowances which have been established in modern archaeoastronomical studies (e.g., Hawkins 1968 and Thom 1967 and 1971), the tolerance of a 15°-20° error will appear absurd. However, after having lived in Misminay for a number of months, I began to understand the equal absurdity of giving even everyday directions by referring to relatively unimportant places or insignificant landmarks on the horizon. This principle applies equally well when relating astronomical phenomena to terrestrial landmarks.

The principal significance of the above data is not only that the informant terminologically equates two opposing groups of stars, but also that the two celestial Collcas are related to his ordering of terrestrial space. In effect, this establishes a diagonal "Collca axis" such that a celestial and terrestrial unit of space in the *northeast* is identified with another in the *southwest*. The two intercardinal units provide reference points—and an axial orientation—for ordering horizontal (terrestrial) space and cyclical (celestial) space and time from the point of view of Misminay. A similar kind of diagonal orientation for an important celestial/terrestrial axis is found in Sonqo. In this case, however, the celestial referents are solar rather than stellar: Apu Ausangate and the rise of the December solstice sun are related as a unit in opposition to Apu Pitusiray and the set of the June solstice sun.

Simultaneous observations of the Pleiades and the tail of Scorpio have been noted elsewhere in the Andes. In the central Andean community of Sarhua (Department of Ayacucho), Dr. John Earls has found that New Year's Day is celebrated on the morning of the winter solstice, June 21, when the solstice sunrise is observed in conjunction with the heliacal rise of the Pleiades and the heliacal set of λ Scorpii (personal communication, 1975; also see Morote Best's discussion [1955] of the relationship of the Pleiades and the June solstice).

In Misminay, the Pleiades are also observed on the morning of June 24 for making crop and weather predictions for the coming year; in effect, June 24 can be considered as an "agricultural" New Year's Day. In both Misminay and Sarhua, then, the June solstice sun is observed and conceptualized in terms of a northeast-southwest axis defined by the rise of the Pleiades and the set of the tail of Scorpio. We should now recall the discussion in chapter 3 of the southeast-northwest diagonal axis based on the conjunction of the December solstice sunrise with the dark cloud constellation of the Fox (Atoq) and the birth of foxes in the antisolstitial (i.e., the June-solstice sunset) direction. These two solstitial axes are diagramed in figure 35. It will be noted that the orientations of the two axes are close to the orientations of the two footpath/irrigation-canal axes (fig. 10) which divide Misminay into four quarters. These data not only support our earlier hypothesis that the two diagonal axes in Misminay are conceptually extended to the horizon where they intersect with the four points of the solstices, but they also support our more general observations about the intercardinal orientation of the divisions of space in Misminay.

From the data on the two Collcas, we see that the terminological equation

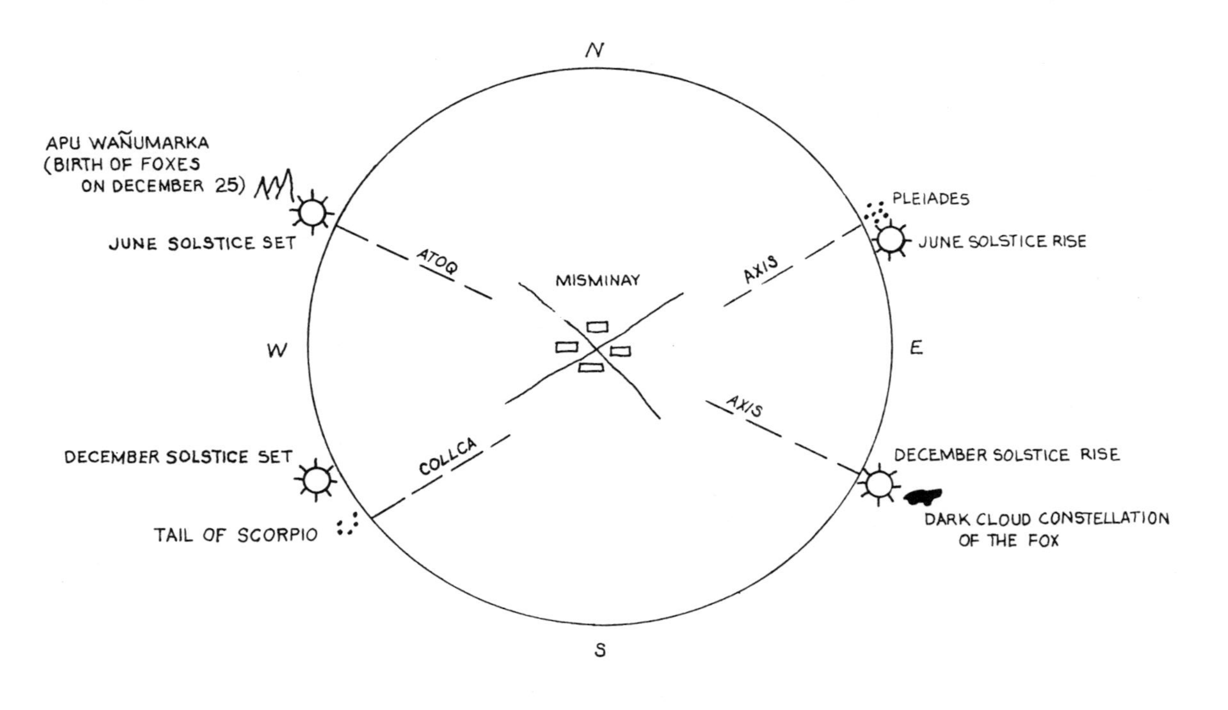

Fig. 35. The Collca and Atoq Axes

of the tail of Scorpio and the Pleiades is more than a simple formalization of the observable fact that the two sets of stars are in close celestial opposition. Temporally, the opposition divides the sky into roughly symmetrical sections so that time at night can be calculated by the position, with respect to the horizon and the zenith, of either of the two Collcas. Spatially, the two Collcas, along with their respective terrestrial associations, provide an axis which can be used as a line of orientation for calculating distances in celestial and terrestrial space. The two Storehouses therefore are important elements in the structure and organization of cosmological orientations. However, their utility is not confined solely to the ordering of cosmological time and space.

The Agricultural Role of Collca (the Pleiades)

Collca is the most consistently observed stellar phenomenon for agricultural purposes. One informant in Misminay partitioned off a section of the night sky from Sirius to the Pleiades and said that this part of the sky is related to maize agriculture. For the moment, then, it is necessary to make a distinction between the two Collcas; the following data refer primarily to the Pleiades.

In an agricultural context the Pleiades are used primarily to help determine when planting should commence and as a preplanting indication of how good the coming year's crop will be. However, the actual methods of agricultural prediction are more complex and varied than this statement suggests. The agricultural observations of the Pleiades can be divided into two categories depending on whether they are made once a year at the beginning of the agricultural season or a number of times throughout the course of the planting season.[3]

The Single Collca Observation. The observation is made for the dual purpose of providing a long-range forecast of how good the crops will be and for determining (relatively) when planting should begin. This initial annual observation in Misminay is made on the morning of June 24, at the end of the all-night celebration of the festival of San Juan. It is made, therefore, after the heliacal rise of the Pleiades in the area of Cuzco (June 3) but a month or two before the actual planting begins. In a number of communities in the Cuzco area, crop predictions made by observing the Pleiades occur not only on the day of San Juan (June 24) but also on the first day of August (Lira 1946:19 and Mishkin 1940:240). As noted earlier (see fig. 3 and table 5), the period around the beginning of August is important in Misminay in relation to the celebration of certain saints' days and as the time when the sunrise is observed to determine the beginning of the planting of maize.

June 24 is also the time when other celestial and noncelestial crop predictions often take place. Near Sicuani, Benjamin Orlove (1979:91) found the following method of prognostication, based on the observation of the June-solstice sunrise:

"There are three possible ways for the sun to appear: *ch'usa* (small) which is the sign of a bad year, *ransa* (large), which predicts a normal year, and *qhoto* (green), which forecasts a very good year." In addition to the translation of *qhoto* as "green," given by Orlove, it seems reasonable to suppose that this term as used in Sicuani could refer to the practice of simultaneously observing the solstice sunrise and a star or group of stars; as we have seen, the cluster of the Pleiades is referred to as *qutu* in the area of Cuzco. As for noncelestial prognostications, we find the following practice recorded in the community of Sicaya (Department of Junín): On the evening of June 23, one peeled potato and two unpeeled ones are thrown under the bed. On the following morning, the person probes around under the bed without looking and grabs one of the three potatoes. If the peeled potato comes up, the next year's crop will be bad; if one of the two unpeeled potatoes is grabbed, the crop will be good (Escobar 1973: 101).

In one variation of the annual Pleiades observational method in Misminay, it is said that when the Pleiades (Collca, Qutu) rise very large and bright, the crops will be abundant; when they rise small and dim, the crops will be poor. The correlation of a large celestial body with good fortune and a small one with bad fortune is similar to Orlove's description of the prognostications based on observing the June-solstice sunrise. The above method of prediction or prognostication based on observing the Pleiades is also the same as that reported by Francisco de Avila some four hundred years ago for the Pleiades (Las Cabrillas) in the central Andean community of Huarochirí (1966:chap. 29). In order to understand the second variation of the single annual observational technique, we must first discuss an unusual stellar phenomenon in Quechua astronomy.

In all four communities in which data were collected, one of the most consistent bits of astronomical lore was that Collca is observed at the time of planting not only to see whether or not *all* of the stars of Collca are bright or dim, but to note the position of the "brightest" star of the cluster. At this time of the year (the planting season), one star in the group is said to appear much larger and brighter than the others. The prognostication of the best time to plant is done, in general, by determining the position of the brightest star relative to the other stars of Collca—whether it appears at the head, in the center, or at the rear of the cluster. This principle is utilized in the single observational method in the following manner: when the brightest star of Collca is located at the rear of the group, this means that planting should not begin until late in the season (October-November); when it rises at the head of the cluster, planting may begin earlier (late July-September). In the community of Kauri, Bernard Mishkin (1940:240) also found that the Pleiades are observed on the day of San Juan and on the first day of August "to see whether the forward, center or rear stars are most brilliant. The position of the most brilliant stars will determine whether there will be an early, middle or late sowing."

Multiple Collca Observations. In addition to the single observation of the Pleiades at the end of June, a number of other observations take place throughout the planting season. One type of multiple observation is based on the principle that Collca must be large and bright before the planting of *each* crop. That is, after it has been determined that the general time to plant has arrived (this on the basis of climatic conditions, the amount of rainfall, the position of sunrise, and divination with coca leaves), planting should not actually begin unless Collca is large and bright the night before each crop is sown.

In a variation of the multiple observation of Collca, the principle of determining the relative position of the one bright star is again important. If the brightest star appears at the back of the Collca cluster it is too early to plant; if it appears at the center it is time to plant; and if at the front it is too late to plant. Most noteworthy in these data is that the "star" appears to be moving with respect to the Pleiades.

The informant who described this last technique was perhaps not the most qualified to testify about the observations of Collca made for agricultural purposes.[4] However, the description is included because it agrees with the data on the single annual Collca observation. As a general principle, the two "bright star" observational techniques suggest that the brightest star of Collca can appear in different positions from year to year, that it is in motion with respect to the other stars of Collca, and that its appearance at the back of Collca means that planting should begin later (back) in time and its appearance at the front means that planting should begin early.

The brightness of the Pleiades cluster, then, is basic to the methods of Collca observation made for agricultural purposes, whether these are single or multiple observations. If the Pleiades is large and bright, this is a positive sign (positive both for planting and for the eventual size of the harvest); if it is small and dim this is a negative sign. Seemingly, there is thought to be some form of direct connection or relationship between the Pleiades and the crops. In this regard, note that the cycle of visibility and invisibility of the Pleiades is closely coincident, respectively, with the presence and absence of maize in the ground. The relationship is outlined in table 8.

Maize is planted and in the ground for approximately the same length and period of time that the Pleiades are visible in the sky. Possibly, this is one reason for directly relating Collca to agriculture, especially to maize agriculture (the relationship has also been pointed out by T. S. Barthel [1971:108]). It should be mentioned that although the true heliacal rise date of the Pleiades at the latitude of Misminay occurs on June 3, I was told explicitly by one informant that Collca "appears" (= heliacal rise?) on June 13, not on the third.

Table 8. The Cycles of Maize and the Pleiades

Maize		The Pleiades	
planting	mid-July	heliacal rise	June 3
harvest	mid-May	heliacal set	April 18
absence	May 15-July 15	invisibility	April 18-June 3
period of absence	ca. 60 days	period of absence	ca. 45 days

The Mobile "Star" of Collca

Two of the four Collca observations just discussed involve a bright star which, from year to year, appears in different positions with respect to the Pleiades. Because any movement of the stars *relative to each other* takes place only over very long periods of time, we can immediately rule out the possibility that the bright, mobile "star" of Collca is a star at all. It should be emphasized that although Collca itself is said to vary from bright to dim, this is never said to be the case with the mobile star; it is always bright, always in motion. Data similar to these from Misminay have been reported elsewhere in the literature. In discussing the August crop predictions made in the Cuzco area, Lira (1946:18) says that a group of stars called *kkóto* (= *qutu*, cf. *SC#* 32) is observed to see "if kkóto has mostly small or large stars, and if one of them moves to another place" (see also the above quote from Mishkin, pp. 119-120).

The celestial latitude of the Pleiades is some 25° north of the equator, which means that it is a few degrees north of the ecliptic (= the path of the sun, the moon, and the planets through the stars). In fact, the ecliptic cuts through the center of the 10° of space which separates the Pleiades from the Hyades. Thus, the path of the planets through the stars will regularly bring them near to the Pleiades. Therefore, it is possible for planets to be seen in front of, at the rear of, and in the "middle" of Collca (i.e., *between* the Pleiades and the Hyades), and the mobile "star" of Collca could actually refer to the planets when they are seen near the Pleiades. This is not to say that a planet will always be in the vicinity of the Pleiades, especially on a single day of the year, such as the morning of the June solstice or the first day of August. Until more data are available, I would make the following hypothesis only for a large area of space (e.g., 15° in front of and behind the Pleiades) and for a relatively long period of time (from the June solstice to the first of August).

On June 24, at the end of the festival of San Juan, the Pleiades appear about 37.5° above the eastern horizon at sunrise; on this day, the Pleiades rise about two and one-half hours before the sun. Thus, the Pleiades, and any planet which happens to be moving along the ecliptic in its vicinity, will be seen in the early morning sky for an hour or so before the rise of the June solstice sun. In the community of Misminay, both Collca and the morning star, *pachapacariq ch'aska* ("dawn of the earth/time star"; *SC#* 9), are observed on the morning of June 24. As the morning star rises, the following song is sung:

Pachapacariq ch'aska locero,	Morning star, day-bringer,
Pachapacariq ch'aska locero.	Morning star, day-bringer.
Uskacha wakicha illarikusun,	Let's shine at once, both of us,
Uskacha wakicha illarikusun.	Let's shine at once, both of us.
Uskaña wakiña illarikusun,	At once–both of us,
Uskaña wakiña illarikusun.	At once–both of us.
Illarisuncha pakarisuncha,	Shine and start the day,
Illarisuncha pakarisuncha.	Shine and start the day.

The informants from whom I recorded this song all insisted that it should be sung in the early morning hours of June 24. This is the day in Misminay when agricultural predictions are made for the coming year by observing the Pleiades, and that fact suggests that the bright star which moves through Collca is the morning star, pachapacariq ch'aska.

On the morning of June 24, 1976, the morning star was Jupiter. It was located about 6°30′ *ahead* of the Pleiades. This position, according to the methods of prediction discussed above, indicated that planting should begin "early" in the season, between July and September. In 1976, the planting of the maize crop in Misminay did, in fact, begin on September 23.

To this point only the stellar and planetary requirements for determining when to plant have been considered. As shown in chapter 3, the phases of the moon are also related to the planting season and looking again at this relationship is an indirect way of arriving at the agricultural importance of the tail of Scorpio.

The Planting Moon and the Two Collcas

An informant in Yucay, an old man who has planted maize in the Urubamba Valley for some thirty years, said that maize should be planted during the waxing moon; that is, from the new moon to the full moon (the period of decreasing wañu). One of the most elementary principles of observational astronomy is that the full moon always appears exactly opposite the sun. Therefore, when a full moon is in the sky, the sun will be found 180° away. As we have seen, the Pleiades and the tail of Scorpio are not in exact opposition to each other, not exactly twelve hours apart. The tail of Scorpio moves through the sky about 10 hours 30 minutes ahead of the Pleiades. This means that, if a full moon were to set at the same time as the tail of Scorpio, the sun would be located about one and one-half to two hours behind the Pleiades (10 hours 30 minutes + 1 hour 30 minutes = 12 hours). This relationship is diagramed in figure 36.

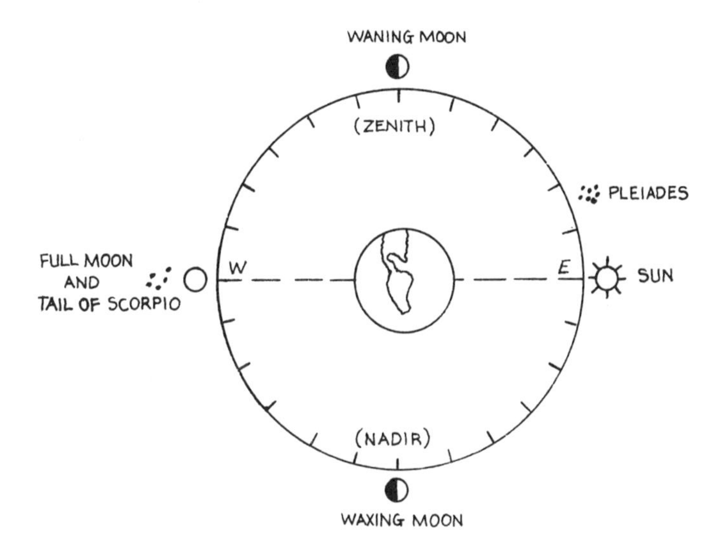

Fig. 36. The Tail of Scorpio and the Full Moon

The earth is at the center of the drawing and the celestial vault stretches around it in a circle divided into twenty-four hours. The figure is drawn so that the full moon and the tail of Scorpio are setting just at sunrise. Since the sun is twelve hours from the full moon and the tail of Scorpio, and since the tail of Scorpio and the Pleiades are separated by 10 hours 30 minutes, the sun is located 1 hour 30 minutes behind the Pleiades.

In figure 37, the drawing is adjusted to show the positions of the Pleiades and the tail of Scorpio as they actually are at sunrise on June 24 (fig. 37a) and on July 24 (fig. 37b). In figure 37a, we see that *if* there is a full moon in the sky on the morning of June 24 (the morning of the observation of the Pleiades) it will set about 30 to 45 minutes after the setting of the tail of Scorpio. There will by no means always be a full moon on June 24, but *every time* there is a full moon on this day it will set at about the same time as the set of the tail of Scorpio and the rise of the solstice sun.[5]

The same celestial bodies appear in figure 37b rotated to their positions on the morning of July 24. This is about the earliest time that planting is actually

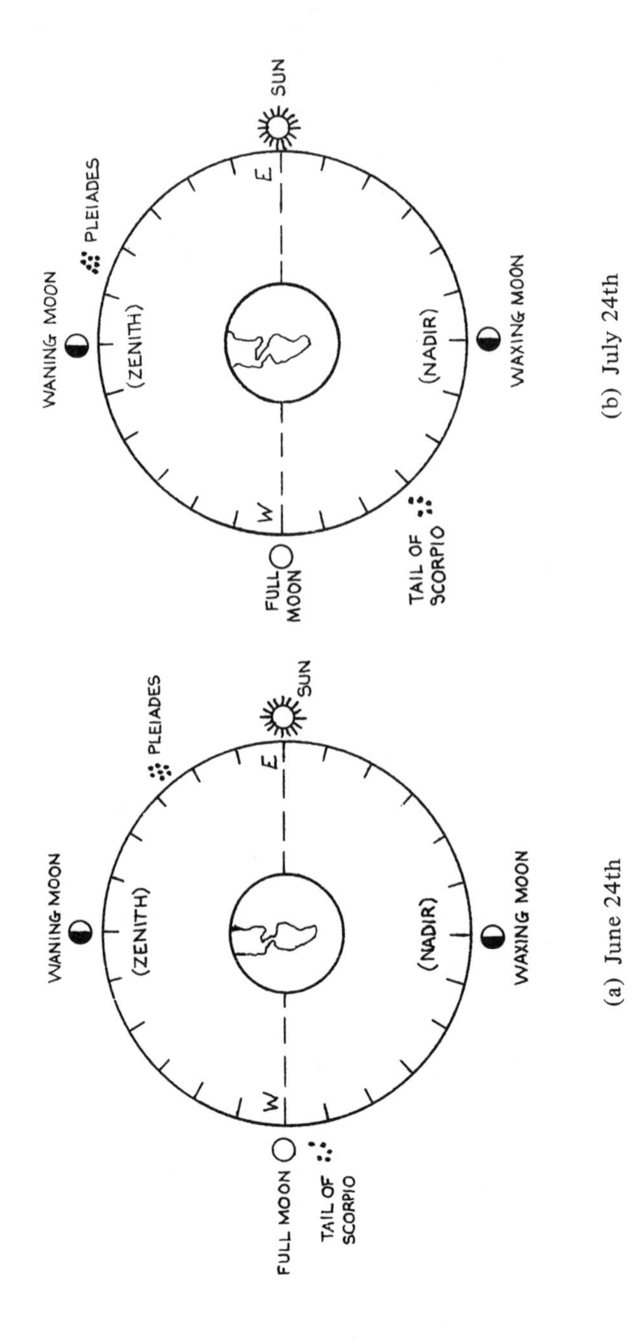

Fig. 37. Planting Moons and the Tail of Scorpio

said to begin and is only a few days before the crop predictions of August 1. At this time, if there is a full moon in the sky, it will set about 3 hours 15 minutes after the setting of the tail of Scorpio. However, if the moon does set at the same time as the tail of Scorpio, it will *always* be the *waxing moon* (halfway between the first-quarter moon and the full moon).

Following these positional relationships through August 24—a time when the planting of maize is more likely to begin—we find that the tail of Scorpio sets within an hour and 15 minutes of a first-quarter waxing moon. It will be recalled that an informant in Yucay said explicitly that maize planting should begin at the time of the first-quarter waxing moon and that it should end with the full moon.

Therefore, if informants say both that planting may begin in late July and that planting should take place during the waxing moon, then *the waxing moon of the planting season will always be seen in association with the tail of Scorpio.* Thus, the agricultural role of the tail of Scorpio, as one celestial Storehouse, would seem to be related to the fact that it will always be seen at night near the planting (waxing) moon, whereas the other Collca (the Pleiades) is seen in relation to the sun.

Collca in the Community of Amaru

Primarily, data from my own fieldwork have been used to develop the argument to this point. However, certain pertinent ethnoastronomical data relating to Collca were collected in the community of Amaru.

Amaru, a community of about 475 inhabitants, is located in the mountains above Pisac, some 35 kilometers northeast of Cuzco (see map 1). The community was studied by Jorge W. Bonett Yépez (1970) and Juan V. Tuero Villa (1973). According to the two ethnographers, crop predictions in Amaru are made three times a year: on June 24, on an unspecified date in July, and on August 1 (Tuero Villa 1973:57 and 94). The predictions of June 24 are made in the morning; those of August 1 are made sometime during the daytime (e.g., rain during the day of August 1 foretells a good year for the crops). The predictions made on the unspecified date in July are the most interesting for the purposes here because they involve both morning and nighttime observations. The early-morning July predictions involve observations of the relative brightness of the constellation Collca.[6]

Concerning the nighttime observations, we read that "during the month of July, the *campesinos* of Amaru observe the configurations within the Milky Way during the night and they predict optimistically or pessimistically about the produce of the year's crop; they do this according to the arrangement of the stars; thus, if groups of stars form the figure of a plow, the crops will be abundant and of good quality" (Bonett 1970:71; my translation). According to Tuero Villa (1973:76-77), "during the evenings of the planting period, constellations should appear [in the Milky Way] in the form of a llama and a plow; these are a very encouraging sign that the next harvest will be good" (my translation).

Although these data do not give a precise identification of the constellation referred to as the "plow" (*arado; SC#* 12), they do support an hypothesis that the plow may be equivalent to the Western constellation of Scorpio. In the quotation from Tuero Villa, we find an explicit association of the Milky Way, the celestial Llama, and the constellation of the Plow. The dark cloud constellation of the Llama (*SC#* 40) is located in the southern Milky Way very near the constellation Scorpio (see fig. 33). Because the Llama and the Plow are observed in Amaru both on the *same* night, we may suppose that they are near each other rather than in opposition.

The shape of the object referred to as an *arado* is important at this point. The plow pulled by two bulls has a fairly standardized form throughout the Department of Cuzco: a detachable yoke and a long sturdy pole connecting the yoke to a curved plowhead and a detachable metal plowshare (fig. 38a). It is very similar in shape to Scorpio (fig. 38b).

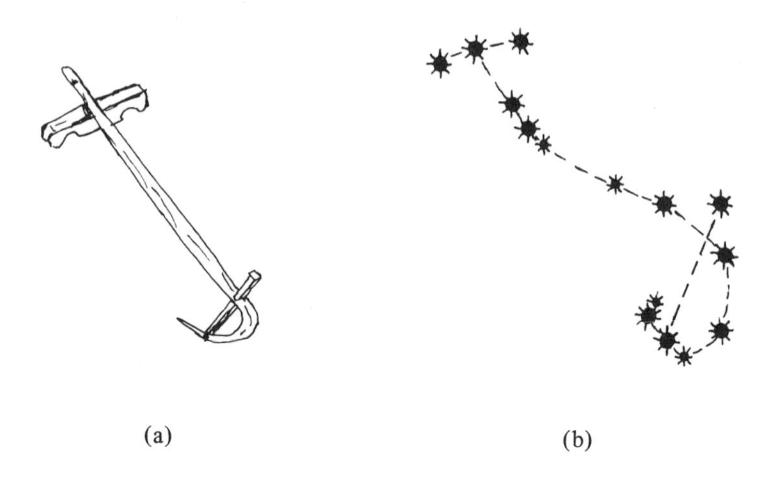

Fig. 38. Scorpio—the Celestial Plow

If this identification for the constellation of the Plow in Amaru is correct, it supports the hypothesis presented earlier concerning the importance of nighttime observations of the tail of Scorpio (Collca) during the planting season. In addition, these data expand the set of elements associated with observations of the celestial Storehouses to include the Milky Way.

In summary, the two constellations referred to as *Collca* are the principal star groups used to time the planting of crops and to make predictions about the size and quality of the harvest. The opposition of the two Collcas is important in the organization of time and space in Misminay and this organization is in accordance with the pattern of intercardinal orientations developed to this point. Finally, the celestial Storehouses are important in the calendar of Misminay because they each integrate a number of terrestrial and celestial phenomena with the agricultural cycle (see table 9).

Table 9. The Associations of the Celestial Storehouses

Collca I	Collca II
Pleiades	Tail of Scorpio
Sun (solstice)	Moon (waxing)
Planets	Milky Way
Morning	Evening
Northeast	Southwest

7. Crosses in the Astronomy and Cosmology of Misminay

The analysis of the structure of terrestrial and celestial space (chap. 2) showed that the division of space along two intersecting axes is an important principle of cosmological organization in Misminay. This is seen within the community where two intersecting footpath/irrigation-canal axes provide the framework for the movement of people and the distribution of water; a similar axial/cruciform framework, based upon the intersection of the two axes of the Milky Way, was described for the structure and orientation of celestial space. The points at which these celestial and terrestrial axes intersect are labeled as "crosses" (Crucero = the terrestrial cross; Calvario = the celestial cross). The importance of the intersection or crossing of two axes has also been found in the chapters on solar cosmology and the celestial Storehouses (Collcas). Therefore the cruciform and the principle of intersecting axes are important structural elements in the cosmology of Misminay.

Beyond these examples of the cruciform in cosmology, there are a number of other ways in which crosses serve as important symbols and organizing principles in the life of the community. For instance, three community crosses made of wood are housed within the central chapel of Crucero (all three are in the shape of the "Latin" cross: †). These are arranged with one large cross in the center and a smaller one on either side. Most communities in the Andes have a number of crosses and chapels or both located at various places on the horizon (see Isbell 1977:86 and Palomino Flores 1968). This is true of several communities in the Cuzco region, but is not the case in Misminay. Informants say that a large cross once stood on the summit of Apu Wañumarka at the place called Taytapata; however, this cross is now located in the church in Santa Ana. In addition to such images of the cross in Misminay, the "sign of the cross" is an important part of ritual and ceremonial observances. For example, during the animal-marking ceremonies at the time of Carnaval, the ritual paraphernalia

Table 10. The Forms and Meanings of Quechua Crosses (*Chaka*)

Quechua	Spanish	Translation of the Spanish
Guardia Mayorga (1971):		
chaka	puente cadera de mujer	bridge hips of a woman
chakana	tejido de carrizos o maguey que se cuelga del techo para colocar algo. *Astr.:* Agrupación de las estrellas de la Cruz del Sur (U).	Weaving of reeds or maguey which is suspended from the ceiling in order to arrange something in its place [e.g., a partition]. *Astr.:* the grouping of stars of the Southern Cross.
chakana, chakata	cruz	cross
chakatay	cruzar, crucificar, poner dos cosas atravesadas como los brazos de la cruz.	to cross, to crucify, to place two things across (through) each other like the arms of the cross.
chakatasqa	crucificado	crucifixion
chakay	cruzar palos	cross sticks
Lira (n.d.):		
chaka	puente, umbral, travesaño, tranca, obstáculo interpuesto. Pierna, nalga, muslo, pernil.	bridge, lintel, crossbeam, crossbar, intervening obstacle. Leg, buttock, thigh, leg (of trousers).
González Holguín ([*1608*] *1952*):		
chacatana, curcu	cruz	cross
chacatani o chacani	cruzar, atravesar	to (go) across
catachillay orochillay	cruzero estrellas	cross of stars (Southern Cross)
chaca	umbrales la cadera	threshold, lintel hip
chacana	escalera	stairs, ladder
chaccana	tres estrellas que llaman las tres Marías	three stars which they call the three Marys [= the belt of Orion]
chacatani	cruzar o poner dos cosas cruzadas o através como los dos brazos de la cruz	to cross, or to put two things crossing (over) like the arms of the cross.

of the *misa* (a blanket upon which the ritual objects are laid out) are arranged in a cross at the end of the ceremony and blessed with libations of chicha. At the beginning of the ceremony, a mixture of chicha and a fine red powder is prepared and crosses are painted on the palms and foreheads of the participants.

Many more examples of the use of cruciform symbols and images in Misminay could (and will be) cited, but the above examples are sufficient to convey the importance of the cross in the religious and symbolic life of the community. In addition, crosses are among the most common forms of star-to-star constellations in the astronomical system. By studying various shapes, groupings, and interrelations of the celestial crosses, we can better understand the symbolism of the cross in modern Quechua and pre-Spanish cosmology (see Aveni et al. 1978, Palomino 1968, Quiroga 1942, Urton 1980, and Zuidema 1973).

Before discussing the celestial cruciforms in Misminay, it is important first to establish the range of shapes, terminologies, and concepts in Quechua which are subsumed within the English word *cross*. Table 10 lists such Quechua terms with their Spanish and English translations.

It is apparent that the Quechua term most commonly associated with *cruz* ("cross") is *chaka* and that *chaka* and its variants refer to a number of interrelated concepts including cross, bridge, lintel, ladder, hip, and the like. Because the concern here is with the use of the term *chaka* in an astronomical context, it is interesting to note two stellar identifications in the above lists: (a) Guardia Mayorga, citing but not referencing Urteaga (= U), says that *chakana* is the name of the Southern Cross, and (b) González Holguín says that *chacana* is the three stars called *las tres Marías*, which is the common Spanish term for the belt of Orion.

The one other astronomical reference in the above list is González Holguín's designation of *cruzero estrellas* ("stars of the Southern Cross") as *catachillay, orcochillay.* When we collect the ethnohistorical and linguistic references to the terms *catachillay* and *orcochillay,* we find a curious fact: *chaka* is consistently related to *cruz* but never to *cruzero* (the Southern Cross); on the other hand, *catachillay* and *orcochillay* are related to *cruzero* but never to *cruz* (Zuidema and Urton 1976:63-64):

catachillay	el cruzero estrellas (Ricardo 1586)
cruzero, estrellas	catachillay, orcochillay (Ricardo)
catachillay, o urcuchillay	el cruze (González H. 1608)
cruzero estrellas	catachillay orochillay (González)
Catachillay	Cruzero, estrellas (Torres Rubio 1619)
Catachillay	una estrella nebulosa en la Vía Láceta, o las estrellas sobre la nebulosa (Bertonio 1612)
cruzero	estrellas, unuchilla catachilla (Bertonio)

But

chacana	las tres Marías (Ricardo)
cruz	chacatana crucu (González)
chacanna	tres estrellas que llaman las tres Marías (González)
chacana	escalera (González)
chaca	umbrales, la cadera (González)

From these data, we may infer that *cruz* and *cruzero* are related but still different concepts; and that *chaka* refers primarily to the belt of Orion and to non-Latin forms of the cross while *cruzero* refers to the Latin cross. Therefore, the central cross in the well-known drawing of Pachacuti Yamqui (fig. 39), which is labeled *chacana en general*, refers perhaps not only to the Southern Cross as is often suggested (Isbell 1978, Sharon 1978, and Zuidema 1973), but also as Pachacuti Yamqui says explicitly, to the "general" concept of two lines or axes which form a chaka (bridge, hip, ladder, cross, and so forth).

On the basis of these definitions and linguistic associations, I hypothesize that in pre-Spanish times the word for "cross," or the term which conveyed many of the same formal and symbolic associations as *cruz* does today, was *chaca(na)* ("bridge" or "crossbeam"). That the term *chacana* was used in an astronomical context in Inca times is supported by the following ethnohistoric data: Juan Polo de Ondegardo (1916:5) mentions the term *chacana* in a list of several unidentified constellations; in the cosmological drawing of Pachacuti Yamqui (1950:226), we see a Latin cross in the center of the drawing labeled *chakana en general* (fig. 39); González Holguín (table 10) associates *chaccana* with *las tres Marías* (the Spanish phrase which refers to the belt of Orion); and Pablo José de Arriaga (1920:198) refers to the adoration of a group of stars called *chacras (chacana?)*, which, says Arriaga, are *las tres Marías*. In addition, the term *chacana* seems to have been used for celestial crosses even as late as the 1920s in the area of Cuzco/Misminay. Robert Lehmann-Nitsche (1928) quotes a letter written to him by Luis Ochoa G. dated October 13, 1924. In the letter, Ochoa discusses a celestial cross he found in his fieldwork in the Department of Cuzco:

Cruz grande or verdadero. En algunas poblaciones como Chinchero, le llaman *chuqui* o sea lanza por su figura; en otras le llaman *Jatun Chaca* o Puente Grande (Lehmann-Nitsche 1928:109 n. 2).	*The large or true cross.* In some communities such as Chinchero, they call this figure *chuqui*, or lance; in others they call it *Jatun Chaca* or the Large (Great) Bridge.

The quotation from Ochoa ties together the ethnohistoric data given previously by relating *chacana* to several different "crossbeam" or "bridge" forms: the cross (one example of which may be the Southern Cross), the lance (e.g., the belt of Orion), and the bridge (*chacana*, discussed below). This group of constel-

Fig. 39. The Cosmological Drawing of Pachacuti Yamqui

lation shapes is similar to Zuidema's interpretation of the ayllu/constellation associations in the village of Recuay during colonial times. Zuidema relates the ayllu of the "father" to the belt of Orion and the ayllu of the "mother" to the Southern Cross (Zuidema 1973:fig. 7).

Terminology and Identification of Celestial Crosses

Table 11 contains the names of the celestial crosses I collected in the field. The left-hand column lists the provenience of the account; the right-hand column identifies the cross by reference to the Star Catalogue of table 7.

The only Quechua terms in table 11 are *hatun* ("great"), *huchuy* ("small"), and *yutu* ("tinamou"). The crosses are all referred to by the Spanish term *cruz* ("cross"). Informants agreed that all of the crosses are male. In a serial listing of several constellations, one informant in Misminay affixed the Spanish term *papa* ("father") to each of the star-to-star constellations including the crosses, but he conspicuously omitted it from his naming of the dark cloud constellations. One important observation to be made from the data in table 11 is that crosses are identified in three different ways: (a) as one cross alone; (b) as sets of paired crosses; and (c) as a group of four crosses related to the four "quarters" (suyus).

Single Crosses. A young man in the community of Lucre drew a celestial cross which he called Passon Cruz ("the cross of passion or suffering"). His drawing is reproduced in figure 40a. Figure 40b is a clarification of the original drawing to which the informant added dots to show that Passon Cruz is composed of two intersecting lines of stars.

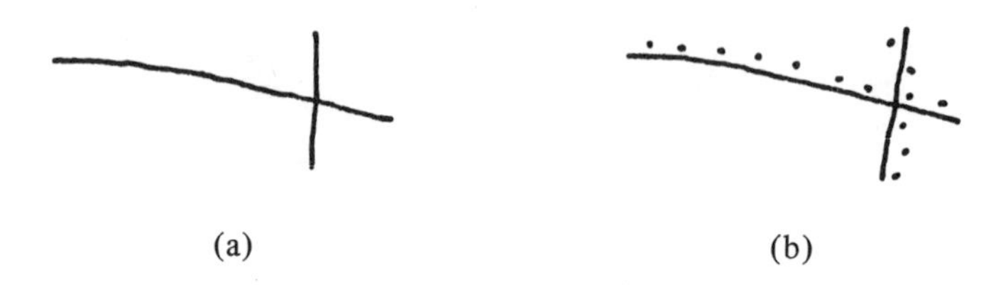

(a) (b)

Fig. 40. The Cross of Passion

An informant from Chumbivilcas identified the Coalsack, a small pear-shaped dark spot in the southern Milky Way, as a "tinamou" (*yutu*), and he then said that the Southern Cross, located next to the Yutu, is Yutucruz ("the tinamou cross").

In Misminay, an informant identified the five stars at the head of Scorpio (α, β, δ, π, and σ Scorpii) as Calvario. Calvario, he said, is *encima de la llama*

Table 11. Celestial Crosses

	Single Crosses	
1. $Misminay_{a*}$	Calvario	*SC#* 15a
2. $Misminay_b$	Calvario	*SC#* ––
3. Lucre	Passon Cruz	*SC#* 29
4. Chumbivilcas	Yutucruz	*SC#* 34
	Paired Crosses	
5. $Sonqo_a$	Huchuy Cruz	*SC#* 21
	Hatun Cruz	*SC#* ––?
6. $Sonqo_b$	Huchuy Cruz	*SC#* 21
	Hatun Cruz	*SC#* 20a
7. $Sonqo_c$	Huchuy Cruz	*SC#* 21
	Hatun Cruz Calvario	*SC#* 20a
8. $Misminay_a$	Linun Cruz	*SC#* ––?
	Calvario Cruz	*SC#* 15a
9. $Misminay_b$	Linun Cruz	*SC#* 25
	Calvario Cruz	*SC#* 15a
10. $Misminay_c$	Linun Cruz	*SC#* 25
	Calvario Cruz/= Papa Dios Cruz	*SC# 28*
11. $Misminay_d$	Linun Cruz	*SC#* 25
	Calvario Cruz/ = Papa Dios Cruz	*SC#* ––?
12. $Misminay_e$	Linun Cruz	*SC#* 25
	Papa Dios Cruz	*SC#* ––?
13. $Misminay_f$	Linun Cruz/= Papa Dios Cruz/= Santíssima Cruz	*SC#* 25, 28
	Cruz Calvario	*SC#* 15c
	Suyu Crosses	
14. $Misminay_a$	Hatun Cruz (north)	*SC#* 20b
	Hatun Cruz (south)	*SC#* ––
	Calvario Cruz (west)	*SC#* 15d
	Calvario Cruz (east)	*SC#* ––

*Subscript letters refer to different informant accounts within the same community.

("above the llama"; the head of Scorpio is located north of the dark cloud constellation of the Llama, *SC#* 40). On another occasion, the same informant identified the celestial Contor (*SC#* 17) as the following stars: *head* = δ, ϵ, η Canis Majoris; *wing* = α Monocerotis; *wing* = α, β, γ Pyxidis; and *tail* = α Hydrae. In a diagram of the Contor, we find that it, too, has the shape of a cross (fig. 41):

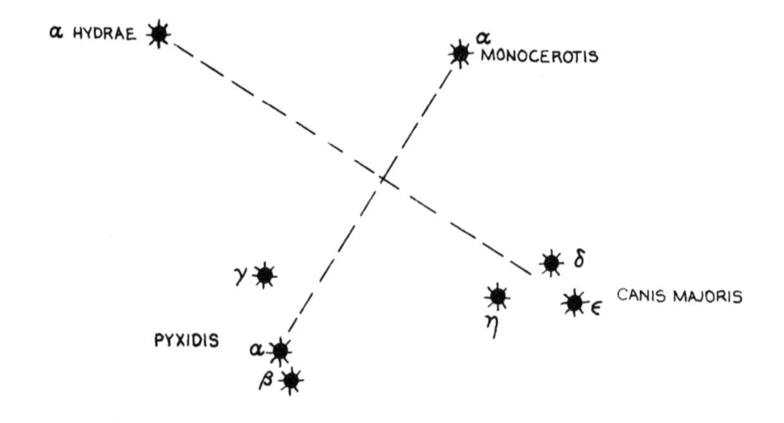

Fig. 41. The Celestial Contor

Another informant in Misminay, a man who later went into considerable descriptive detail concerning the four crosses of the four suyus, also said that there is a single cross called Cruz Calvario and that it is used to determine directions at night. This "directional cross" is shown in figure 42a. In figure 42b I have drawn the same cross as in 42a, but have included the north, south, east, and west points and the azimuths, as viewed from the earth, of the two axes of the "directional cross" when it stands overhead. The directions to which Cruz Calvario points are not the cardinals, but rather the intercardinals.

We should be reminded here that the phrase *Cruz Calvario* is also used in a wider, cosmological context in the community of Misminay. Chapter 2 established the Milky Way as the principal celestial feature used in the orientation of celestial and terrestrial space. The celestial sphere is conceptually divided into four quarters (suyus) by the two alternating, intercardinal axes of the Milky Way as it passes through the zenith (axis: NE-SW → axis: NW-SE). When the Milky Way stands in the zenith, it is called Cruz Calvario. Thus, like the true cross of Calvary in the Christian tradition, which is represented by a cross on the summit of a mountain or one approached by steps (fig. 43), the Quechuas apply the term *calvario* to the celestial cross or crossbeam which moves first from the horizon up to the zenith and then from the zenith down to the horizon.

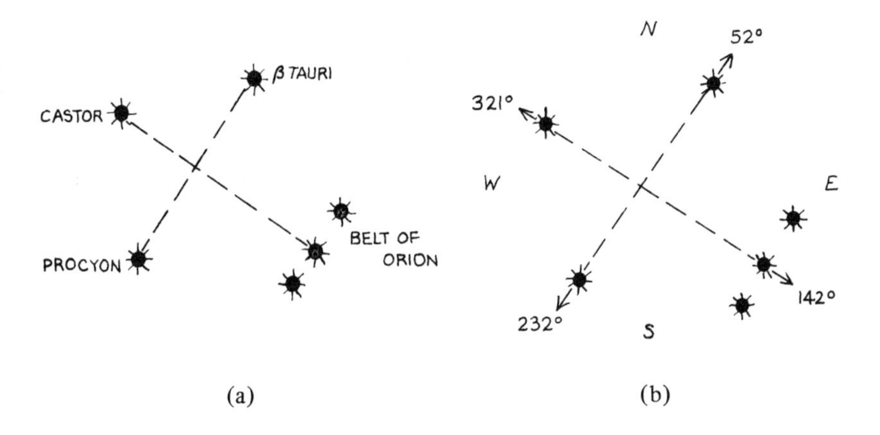

(a) (b)

Fig. 42. The Directional Cross

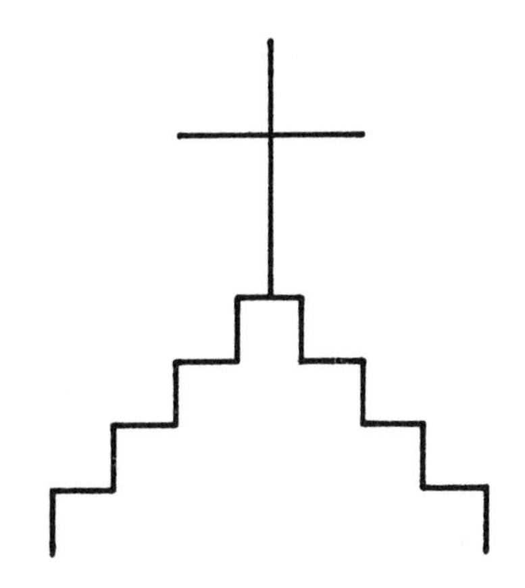

Fig. 43. The Cross of Calvary

Dr. Billie Jean Isbell, in a study of a series of drawings made by school children in the community of Chuschi (Department of Ayacucho), found that the Wamanis, the gods of the mountains, are sometimes equated with a cross which stands on the summit of a mountain. In one figure drawn by a child "the *Wamanis* are represented by mountains with crosses on the summits and with irrigation canals descending from the divine springs" (Isbell 1976:44-46).

The drawing referred to in this quotation depicts an indigenous tradition relating to the Wamanis as the source of water (see Carrión Cachot 1955). Thus, we see a convergence with the Christian iconographic tradition in which the four rivers of Paradise descend either from the foot of the Tree of Life or from the Cross of Calvary (see Ferguson 1904:19 and 58, Schiller 1972, Seymour 1898, and Weckmann 1951:131-132). In a study of the iconography of the cross in Pre-Columbian America, Adán Quiroga (1942:254) concluded that water/rain was the fundamental motive of the different religious systems in the Americas and that its uniform symbol was the cross. This is supported by Isbell's (1970: 90-91) description of the complementary symbolism of crosses and the points of convergence of rivers in the annual irrigation-canal cleaning ceremony (Yarga Aspiy) of the community of Chuschi.

Under the category of single crosses, we should include the celestial identifications directly associated with the term *chaka*. As was concluded above, *chaka* is the principle term used to designate non-Latin cruciforms.

The term *chacana* is employed in a number of contexts in the community of Misminay. In an astronomical context, *chacana* often refers, as it apparently did in Inca times, to the belt of Orion (*SC#* 18). An informant in Misminay described the celestial Chacana as having the shape of a man. The five stars of Chacana represent the five extremities of the body: head and arms = the belt of Orion; one leg = θ Orionis; one leg = η Orionis (see fig. 44). In Sonqo, Chacanuay is identified both as the belt of Orion and as the three stars of δ, ϵ, and η Canis Majoris (fig. 45).

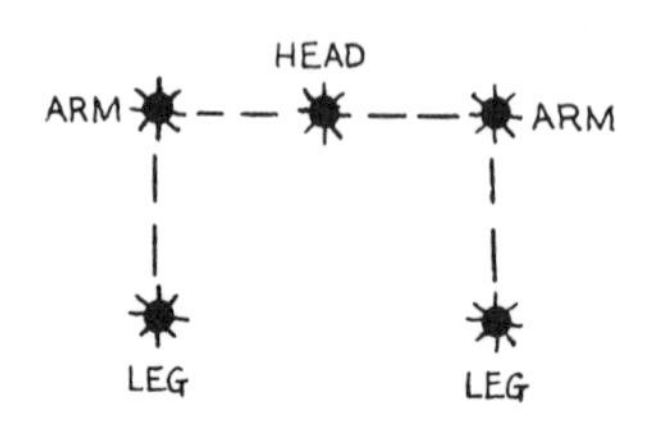

Fig. 44. The Stellar Human/Bridge in Misminay

Fig. 45. Celestial Bridges in Sonqo

The belt of Orion is located along the celestial equator. It is possible that it is considered a "bridge" or "crossbeam" (chacana) because it marks the point at which the sky is divided into two parts; also, its rising and setting points on the horizon coincide with those of the equinox sun. Thus, Chacana, as the belt of Orion, marks the eastern and western points at which the sun makes its transition (crossing) between the northern and southern hemispheres. Since δ, ϵ, and η Canis Majoris are located along the central course of the Milky Way, I suggest that they are classified as chacana because they are thought of as a bridge spanning the celestial River, the Milky Way.

Both of the shapes in fig. 45, the straight line and the open *V*, are seen in the use of the term *chaca(na)* for certain material objects in the community of Misminay. In house construction, the large wooden beams which span a room are called *arma chaca*; perpendicular to these large beams are smaller ones which form the base of the second floor storeroom, the *marka*. These smaller crossbeams are called *marka chaca* (fig. 46).

Fig. 46. House Beams in Misminay

Another use of the term *chacana* concerns the paraphernalia used in preparing chicha. After the germinating maize kernels (*wiñapu*) have been stone-ground into a coarse powder, it is boiled for a few minutes in water. The corn mash is then poured into a large basket (*isanga*) lined with corn husks. The corn husks act as a filter allowing the liquid from the corn mash to drain into a large, open-mouthed urn called a raki. However, the basket does not sit directly on top of the raki; between them is a forked piece of wood called chacana (fig. 47).

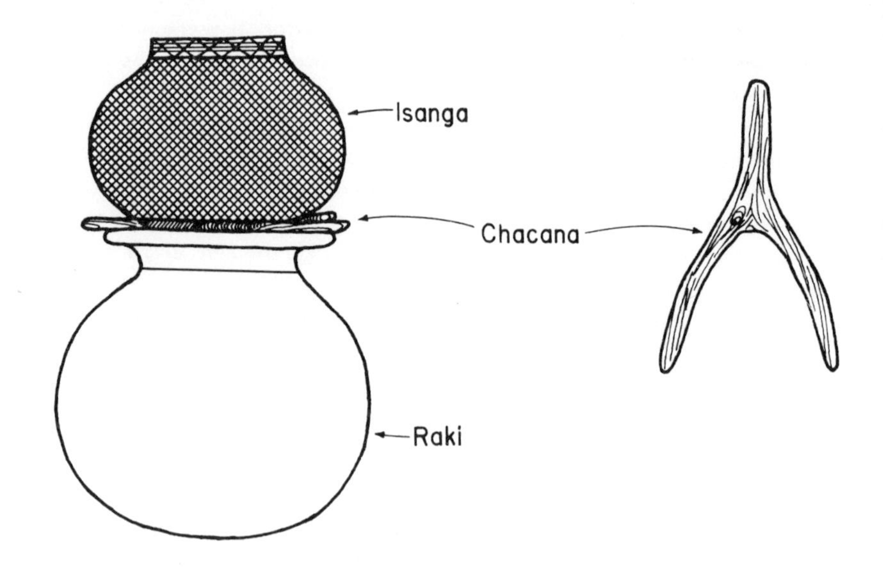

Fig. 47. The Cross in the Preparation of *Chicha*

At first sight, this forked-stick chacana appears to be unrelated to the cruciforms discussed thus far. However, it is perhaps the most "successful" of all in accomplishing the function of the cross (i.e., the bridging of objects and lines of motion) because it performs this function on the horizontal and vertical planes simultaneously. By supporting the isanga atop the raki, it separates two objects in the vertical dimension, thus allowing the chicha to filter from one vessel, and level, to another. Horizontally, when viewed as a two-dimensional form the *Y*-shape represents the convergence of two lines of motion into one. We will see later that this union of opposites (*tinkuy*) is the basic principle underlying *all* forms of chacana as a bridge or crossbeam.

Paired Crosses. In the listing of "paired crosses" in table 11, there is a consistent correspondence between the following crosses:

either: Huchuy Cruz ("small cross") and Hatun Cruz ("large cross")

or: Linun Cruz ("wooden cross")[2] and Calvario Cruz ("Cross of Calvary").

Huchuy Cruz is always coupled with Hatun Cruz (and vice versa) and Linun Cruz is always coupled with Calvario Cruz (and vice versa). The Huchuy/Hatun pair of crosses is found predominantly in Sonqo, the exception being account 14 (table 11), where Hatun Cruz was mentioned in Misminay. Linun/Calvario are found as terms for the paired crosses primarily in Misminay with the exception of account 7 (table 11), where Hatun Cruz Calvario was mentioned in Sonqo.

The specific stellar identifications of these four crosses are somewhat of a problem for there appears to be a great deal of variation, especially in the Linun/Calvario pairing. However, the following identifications are the most probable:

Huchuy Cruz = the Southern Cross
Hatun Cruz = a combination of various stars including Betelgeuse, Rigel, Sirius, Procyon, Castor, η and μ Geminorum.
Linun Cruz = the five stars of the head of Scorpio
Calvario Cruz = some of the same stars identified for Hatun Cruz plus the belt of Orion

Despite the variation in the specific stars associated with any one of the four crosses, there is an overall consistency both in the shape of the paired crosses and in the general celestial area they occupy. Concerning the shape of Hatun Cruz and Huchuy Cruz, both are composed of four stars arranged as though they were the tips of a Latin cross (fig. 48); Linun Cruz and Calvario Cruz, on the other hand, are in the form of *T*-shaped crosses composed of five stars (fig. 49).

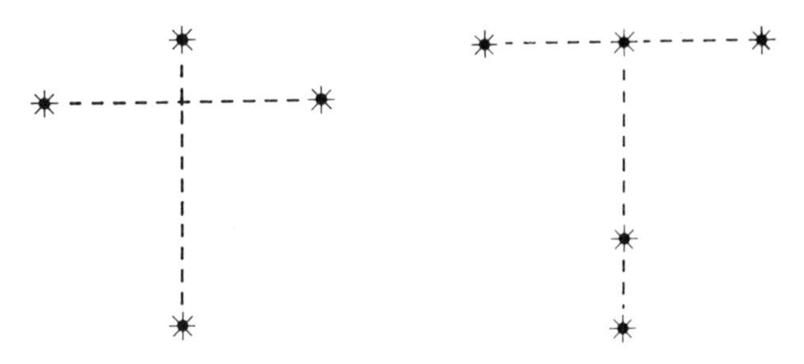

Fig. 48. The Latin Cross

Fig. 49. The *T*-Shaped Cross

In order to demonstrate the consistency in the areas of the sky occupied by the two pairs of crosses, we must refer to a schematic map of the celestial sphere. Figure 50 diagrams the paired crosses as they are related to the path of the ecliptic (the dotted line).

On the basis of figure 50, we can state a few general observations concerning the significance of paired crosses in Quechua astronomy: each of the two members of the paired crosses are roughly in opposition to each other in the sky; Hatun Cruz and Calvario Cruz are located together along the celestial equator and Huchuy Cruz and Linun Cruz are near each other in the south celestial hemisphere; and finally, the paired crosses *may* be opposed to each other in the sky in a way similar to the opposition of the sun among the stars at the times of the solstices (table 12). As the crosses of the suyus, the four celestial crosses maintain the same relation as displayed among the paired crosses (i.e., a relation of solstitial opposition).

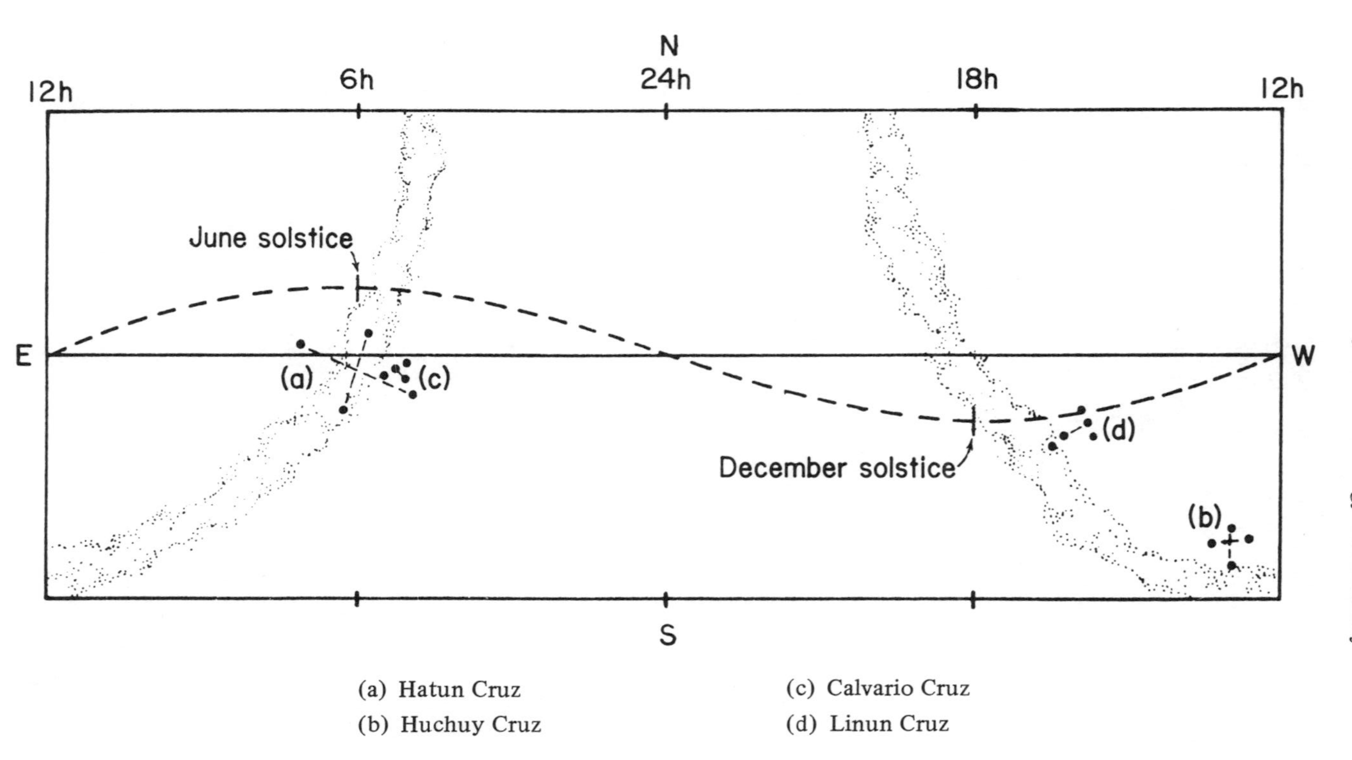

Fig. 50. The Paired Crosses and the Solstices

Table 12. Celestial Crosses and the Solstices

Pair 1		Pair 2		Solstitial Relation
Hatun Cruz	+	Calvario Cruz	=	June Solstice
Huchuy Cruz	+	Linun Cruz	=	December Solstice

The Crosses of the Four Quarters. One of my principal informants in Misminay described an arrangement of four celestial crosses which, he said, relate to the four-part (suyu) organization of the celestial sphere discussed in chapter 2. The four crosses of the suyus are:

Hatun Cruz—"Cross of the North" = Procyon, Castor, η and μ Geminorum
? —"Cross of the South" = α and ϵ Centauri, ρ and σ Lupi
Calvario —"Cross of the West" = belt of Orion, τ Orionis, Rigel
? —"Cross of the East" = $\alpha, \beta, \delta, \pi$, and σ Scorpii

The informant referred to the Cross of the North as Hatun Cruz and to the Cross of the West as Calvario; the other two crosses were not specifically named. A diagram of the four suyu crosses (fig. 51) shows a pairing similar to that described for the two sets of paired crosses. However, it is similar, not identical, because of the introduction of a new shape, a *Y*-shape (the Pall or Forked Cross), which is the shape of the chacana stick used in preparing chicha (fig. 47) and of the absence of the †-shape (the Latin Cross).

Since the Hatun and Calvario suyu crosses are formally different from each other but each is similar in shape to another, unnamed cross, the "crosses of the four suyus" may actually represent a *unification of two sets of paired crosses.* Thus, I suggest that we can associate the unnamed suyu Cross of the South with Huchuy Cruz and the unnamed suyu Cross of the East with Linun Cruz. This unification would represent a transformation in the shape of the Hatun/Huchuy group (fig. 52) from the †-shape of the Hatun/Huchuy crosses discussed earlier to a *Y*-shape (fig. 52). Notice, however, that the Hatun/Huchuy crosses contain four stars in both cases—as paired crosses *and* as two crosses of a four-suyu grouping. Similarly, the Calvario/Linun crosses contain five stars in both cases.

More will be said in a moment concerning the possible relationships of these stellar cruciforms (and their transforms), but first the celestial locations of the suyu crosses must be analyzed. As concluded above, the celestial locations of the paired crosses suggest that they may be related (opposed) in some manner similar to the opposition between the two solstices. Figure 53 shows the approximate locations of the four suyu crosses.

If we compare figures 50 and 53, our observation concerning a possible relationship between the paired celestial crosses and the solstices is supported by

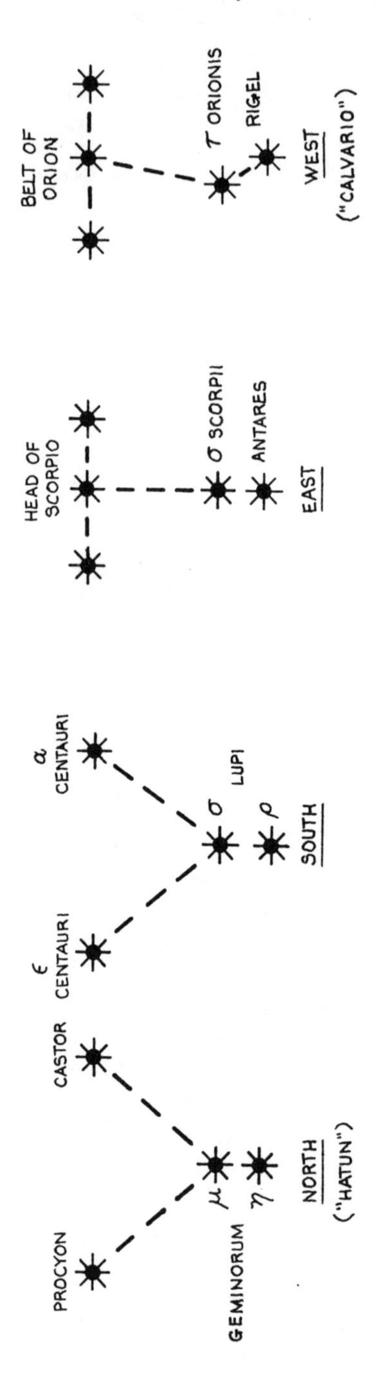

Fig. 51. The Crosses of the Quarters

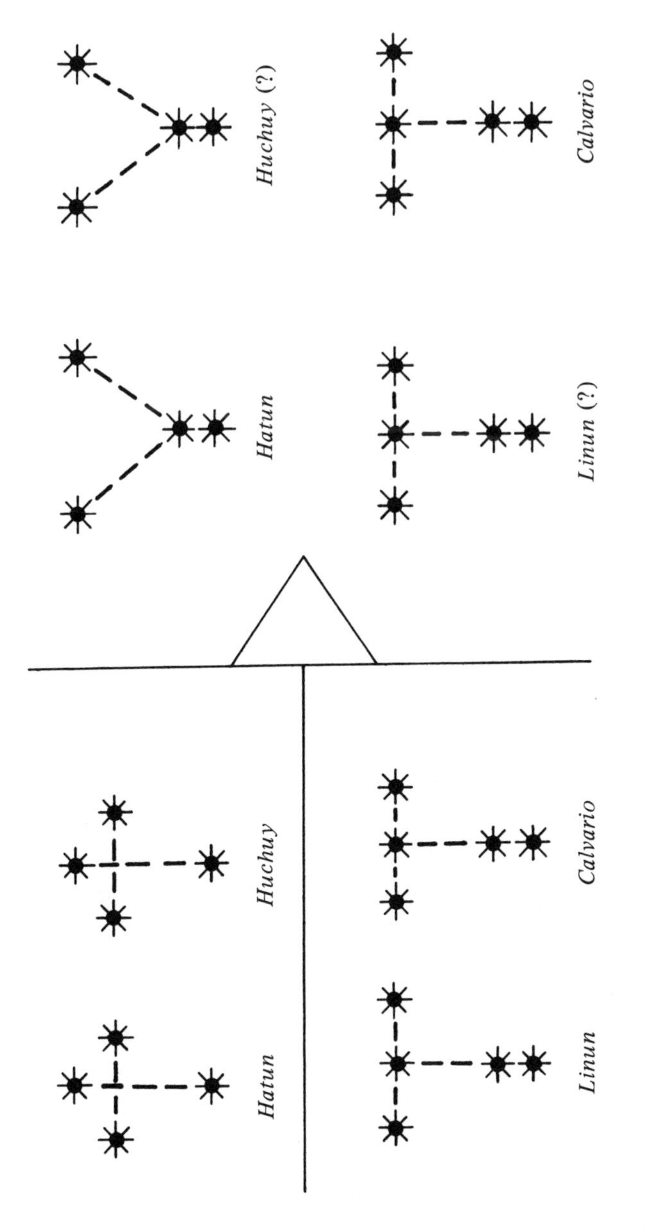

Fig. 52. Paired and *Suyu* Cruciforms

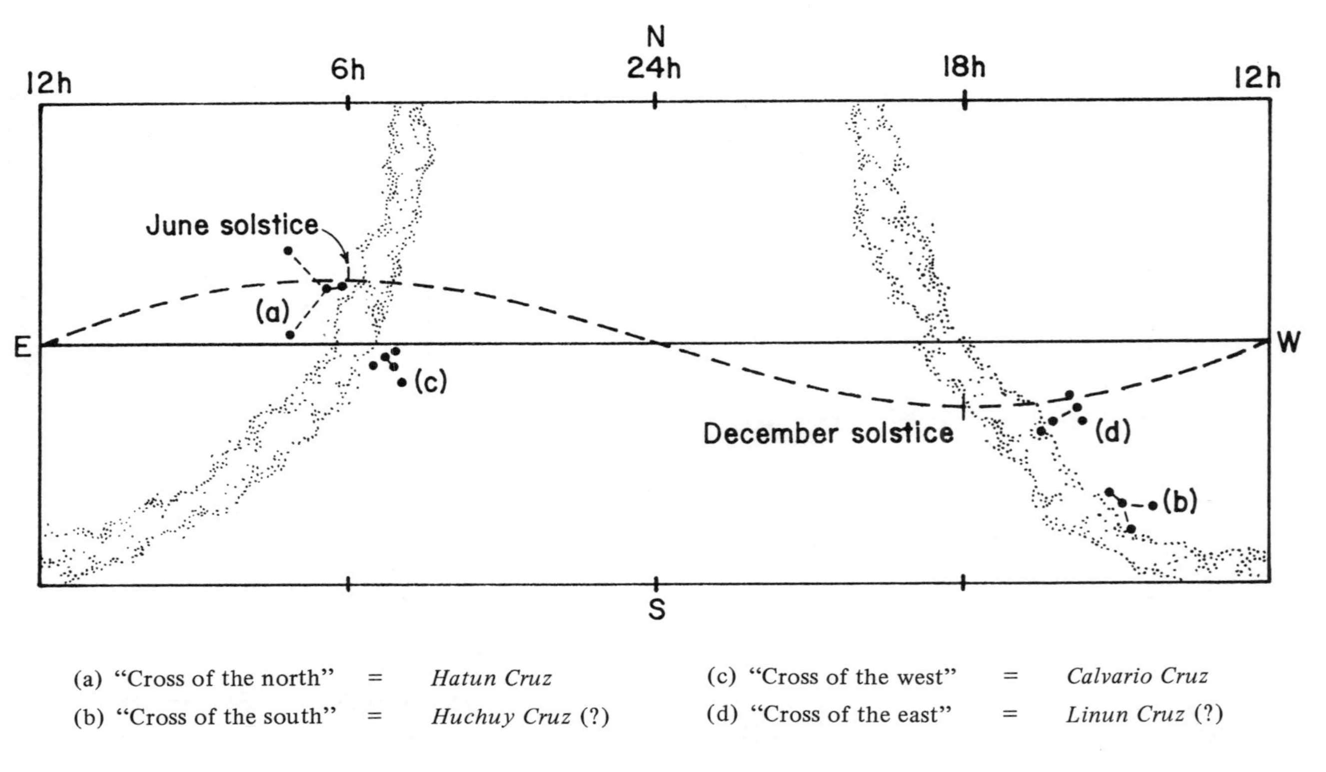

Fig. 53. The *Suyu* Crosses and the Solstices

the data on the four suyu crosses. In addition, we find that the *Y*-shaped crosses in figure 53 (which are transformations of the †-shaped crosses in figure 50) are opposed in the sky in a manner similar to the opposition of the pair of †-shaped crosses in figure 50. The *T*-shaped crosses are identical in both figures. These combined data suggest the following conclusions: (a) the groupings of two and four crosses express an opposition of the north, or the region of the equator, to the south; (b) the opposition of crosses, in groups of either two or four, is an opposition similar, and probably related to, the opposition between the two solstices; and (c) the †-shaped, four-star cross can be transformed into the *Y*-shaped four-star cross, and vice versa.

In addition to these conclusions regarding the structure and organization of the celestial crosses, we should make one more general observation concerning their relative locations. In figure 54, the paired and suyu crosses are combined in a single diagram, which includes the line of the Milky Way as it cuts through the celestial hemispheres. It illustrates a point made repeatedly; the Milky Way is the principal line or plane of orientation in Quechua astronomy. The crosses of the sky are spatially and temporally associated with the solstices. That is, the crosses are clustered along the path of the Milky Way near the two points where the Milky Way is intersected by the ecliptic (the path of the sun, moon, and planets through the stars). This suggests that the celestial crosses may be important in indicating the changing relationship between the sun and the stars at the times of the solstices. As the sun rises in the east with one cluster of crosses, the opposite cluster will set in the west. In future fieldwork, it will be important to collect additional information regarding the calendrical and symbolic importance of the various types of crosses and their relation to the solstices and the Milky Way.

Opposing celestial crosses have been noted elsewhere in the astronomy of South American Indians. In Venezuela, the Warao Indians say that the Northern Cross (Cygnus) and the Southern Cross (Crux) are twin turkey-birds (*Crax alector*) which alternately fly to the meridian and cry for the protection of Warao children (Wilbert 1975:36). The Northern and Southern Crosses are located within the Milky Way, and their transits of the meridian are separated by about twelve hours (see also Reichel-Dolmatoff 1978a, for an excellent discussion of the solstitial/cross symbolism of the Kogi Indians).

The Symbolism of Terrestrial and Celestial Crosses

We are now in a position to address ourselves to the general problem of cruciform symbolism in Quechua. The cruciforms most consistently recognized in the Quechua system of astronomy appear in figure 55.

In the left-hand column, I have arranged the cruciforms in such a way as to

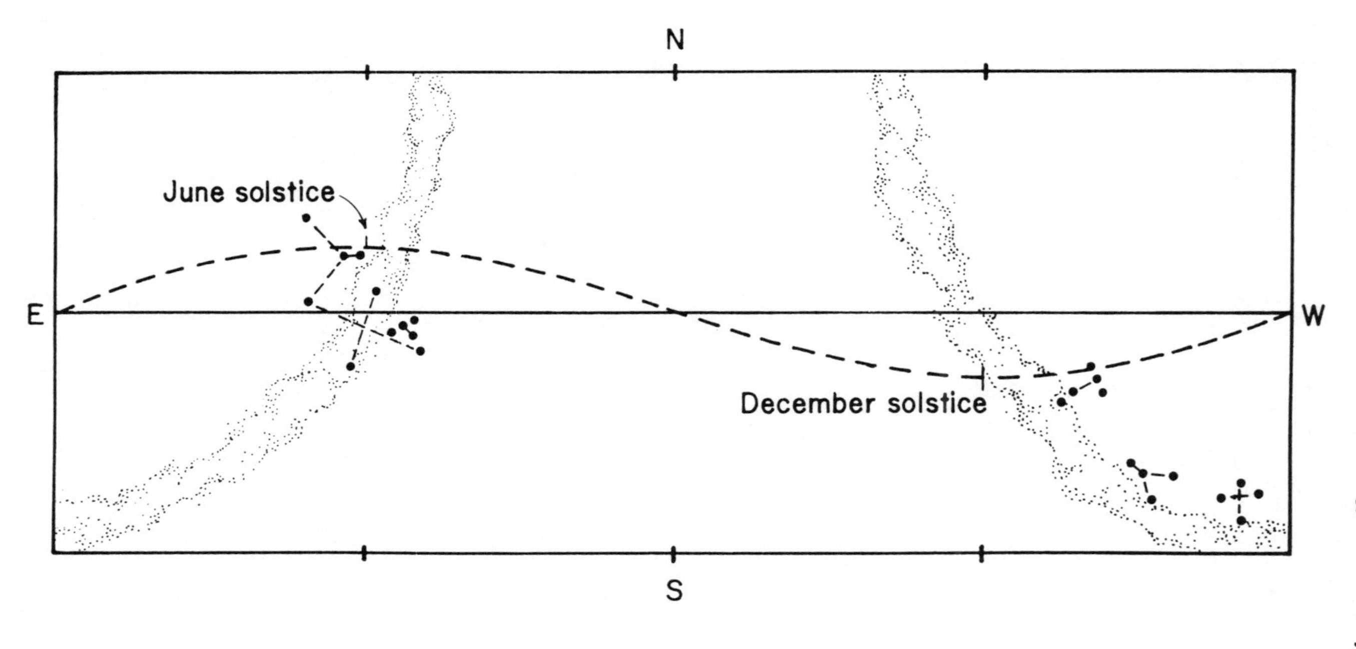

Fig. 54. Crosses, Solstices, and the Milky Way

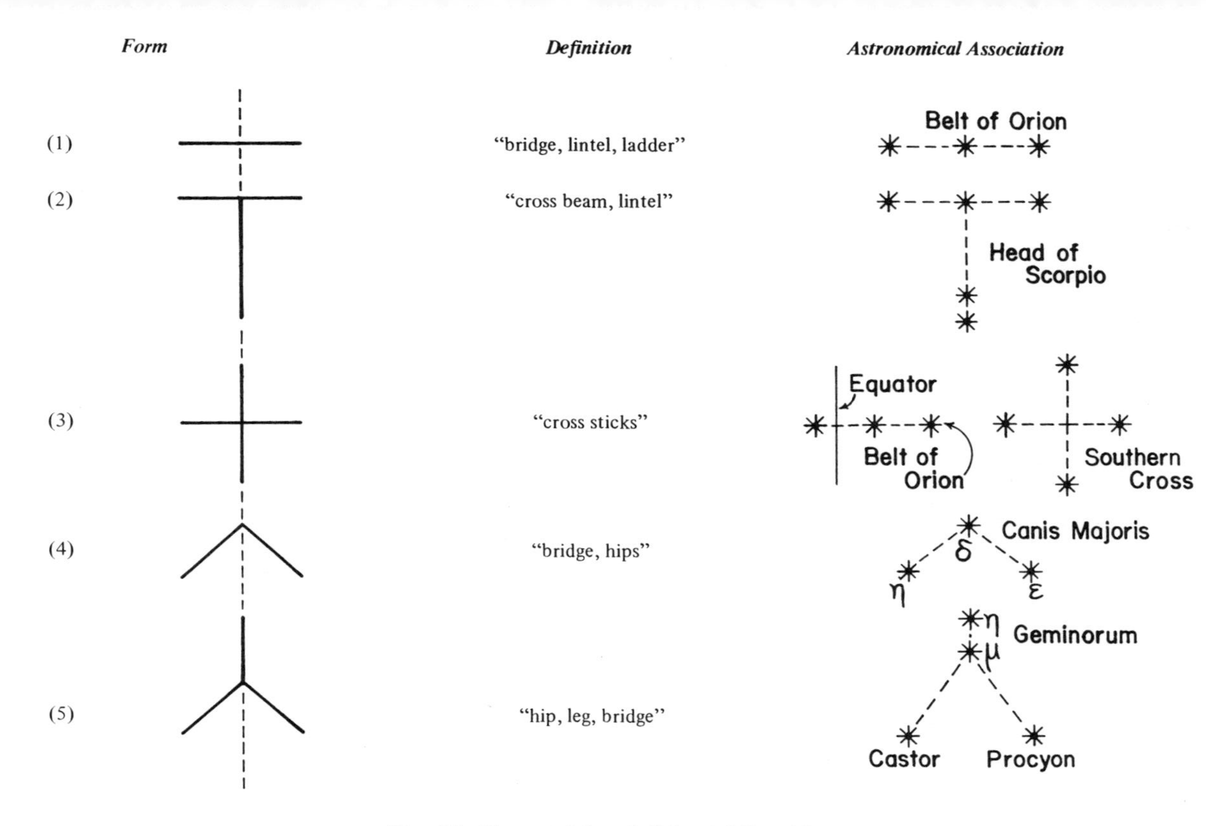

Fig. 55. Terrestrial and Celestial Cruciforms

be able to pass a line (itself a form of chaka) through one of the major axes of each figure. The line illustrates a principle which consistently emerged as one of the central characteristics of all cruciforms discussed above: chaka, whether in the sky or on the earth, is the expression of the principle of simultaneous union/bifurcation. Whether the particular form of chaka serves as a medium for joining two distant points (form 1, above), or for redistributing weight equally down a column (form 2), or unifying two roads or rivers into one (form 5), the underlying principle concerns the balancing of forces and lines of movement, a concept which, in Quechua, is expressed by the term *tinkuy* (see Earls and Silverblatt 1978). In short, chaka is an axis along which a state of equilibrium is established and maintained. I suggest that this is one of the major functions of the concept of chaka in Quechua sociocultural organization and that, as we have seen, chaka (as a manifestation of tinkuy) also serves as one of the principal cosmological notions out of which an ordered view of the universe is crystallized.

It is apparent that the cruciforms in the left-hand column of figure 54 embody the same principle, or set of principles, and that they can be combined in various ways to produce all of the celestial cruciforms described earlier in this chapter. We can also see that the forms, functions, and relationships of the celestial crosses are similar, and probably related, to the cosmological structures and organizational principles described in chapter 2. That is, we have found that the cosmological ordering of space in Misminay is based upon the mirror-image intersection of axes in the celestial and terrestrial spheres. These axes order and coordinate cosmological space and time in a way similar to the ordering of the celestial sphere by the shapes, positions, and interrelationships among the single, paired, and suyu crosses.

8. The Stars of the Twilight

The description of the daily activities in Misminay noted that every day begins with the call of the pichiko bird and the sight of the morning star (*pachapacariq ch'aska* or *illarimi ch'aska*) and that every day ends with the pichiko and the setting of the evening star (ch'issin ch'aska). More importantly, in Misminay, the beginning of primordial time and the very first division between night and day is said to have come about with the rising of the morning star; *pachapacariq ch'aska* means not only "dawn star" but also "dawn of the earth/time star." The following explanation was given by an informant in Misminay.

Dice un mil, dos mil, tres mil no había las estrellas. Entonces, las aparecían; poco a poco aparecia-ban. Ahora, también ya está apareciendo harto. Primer que ha salido es de Pachapacariq; Pachapacariq ch'aska.	It is said that one thousand, two thousand, and three thousand years ago there were no stars; then they began to appear, little by little they began to appear. Now the sky is almost full. The stars appeared little by little and the first to appear was pachapa-cariq ch'aska.

Thus, the first object to rise into the void of the primeval sky was neither the sun nor the moon but the morning star. I was not told why the first appearance of the morning star historically preceded that of the sun or the moon, but it is perhaps related to the notion that the sun could not have risen into a black void; that is, *twilight* preceded the sun. In rising before the sun, pachapacariq ch'aska established, and still establishes each day, the space, time and directionality essential for the sun to pass from the internal or lower world (otra nación) into the upper world. Therefore, the morning star is inseparable from the period of twilight which moves in front of the sun. At the other end of the day, at dusk, the evening star (ch'issin ch'aska) is associated with the space through which the sun has moved during the late afternoon hours; ch'issin ch'aska is visually associated with the twilight of dusk. Clearly, we cannot understand the Quechua conception of the morning star and evening star until we first understand the cosmological significance of "twilight."[1]

Cosmology and the Periods of Twilight

Figure 56 is a diagram of the arcadian cycle. (Certain hour units along the circle are indicated only to provide a method of description and analysis and can be dispensed with later in the description.) In the figure we see that the period from sunset to sunrise is divided into three segments by the sequence:

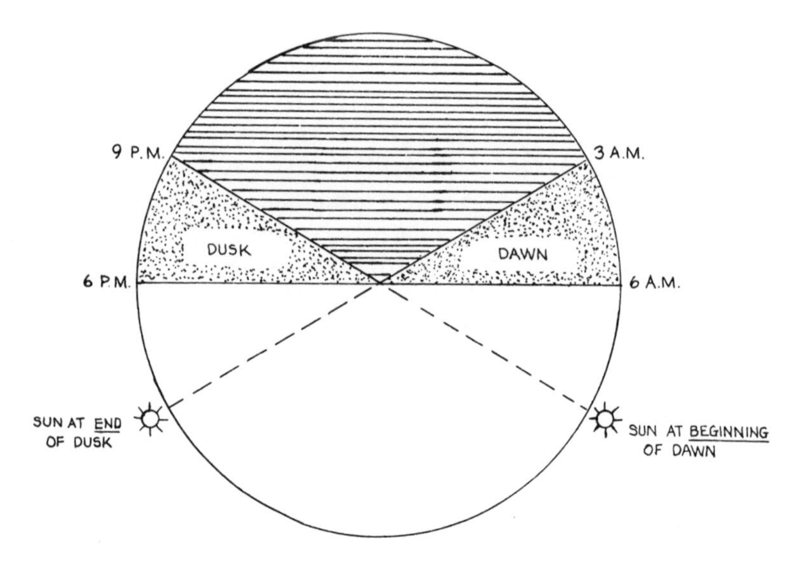

Fig. 56. Twilight in the Arcadian Cycle

dusk → midnight → dawn (which = twilight → dark → twilight). In the full twenty-four-hour period, we find the divisions shown in figure 57.

In figures 56 and 57 note that, for observational purposes, the sun is always preceded by one period of twilight in the morning and followed by another in the evening. This may seem obvious, or insignificant, for we recognize essentially the same divisions and units of time in Western astronomical thought. However, the conception of "twilight" in Quechua cosmology is different from our own, and the difference is based on the fact that the Quechua twilight is not only related to the periods of time but also to two units of space/time which are in continual motion in front of and behind the sun. That is, *pacarin* ("to be

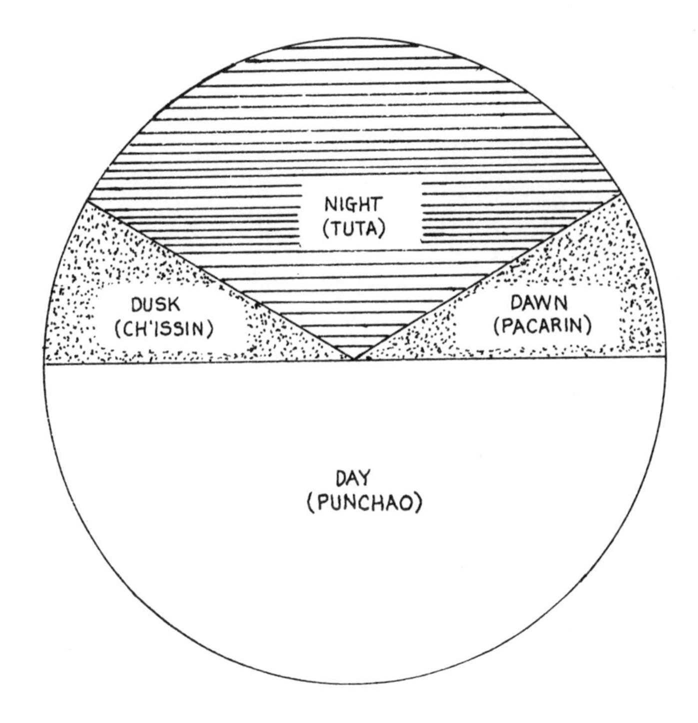

Fig. 57. Periods of the Day and Night

born," "to dawn") is not just the period of time when the eastern sky begins to lighten early in the morning (beginning around 3:00 to 4:00 A.M.); rather it is both this period of *time* and the unit of celestial *space* illuminated by the early morning sun. Any object located within this celestial territory *can be* referred to as pacarin or *illarimi*, both of which mean "dawn," because of their association with the named unit of space/time which precedes the sun. The point is made more forcefully by stating that even when the sun stands along the meridian at noon, pacarin, as a unit of space/time, precedes it but now stands in the *west*. The unit of twilight called *ch'issin* ("dusk") similarly *follows* the sun continuously along the celestial sphere (see fig. 58). Thus, pacarin and ch'issin refer to units of space/time defined *in relation to the sun;* they are the two areas and periods of twilight.

Terminology of the Twilight and Zenith Stars

Table 13 lists all of the accounts I collected in the field regarding the names of the morning and evening stars. In addition, every community where

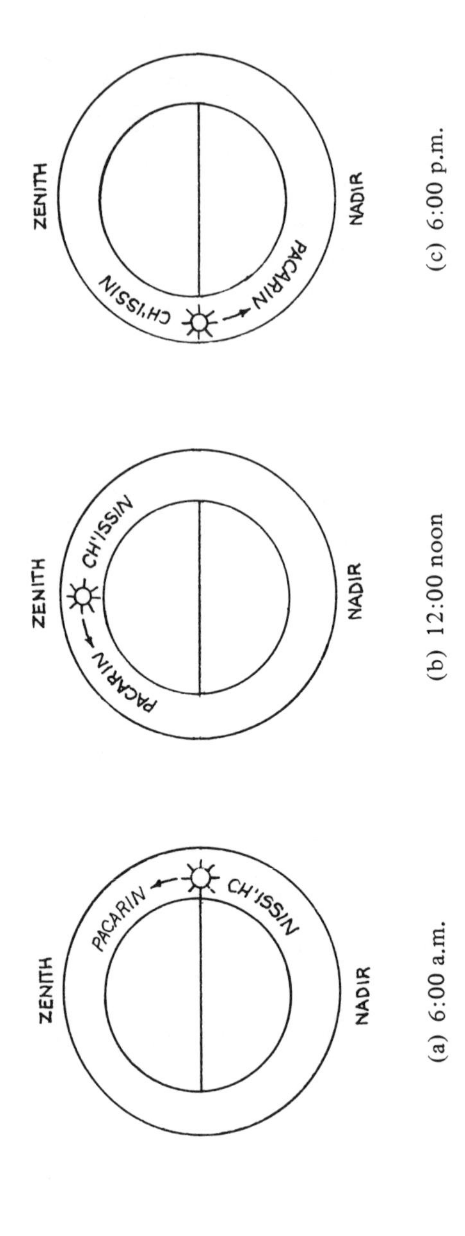

(a) 6:00 a.m. (b) 12:00 noon (c) 6:00 p.m.

Fig. 58. The Revolution of Dawn and Dusk

Table 13. The Twilight and Zenith Stars

Community Where Collected	Evening Star	Midnight ("zenith") Star	Morning Star
$Misminay_a$	ch'issin ch'aska (set ca. 8:00 p.m.)		pachapacariq ch'aska (rise ca. 4:00 a.m.)
$Misminay_b$	______ coyllur	______ coyllur	______ coyllur
$Misminay_c$		Torito	
$Yucay_a$			pachaypaqa locero
$Yucay_b$	ch'issin ch'aska	altopiña karuska ch'issin ch'aska	Illareraimunkana (?)
Quispihuara	ch'issin ch'aska (set ca. 8:00 p.m.)	cuscan tuta	Illarimi ch'aska (rise ca. 4:00 a.m.)
Lucre	locero		Locero (rise ca. 3:00 a.m.)
$Songo_a$	ch'issin ch'aska		pachaillarimi ch'aska
$Songo_b$	ch'issin ch'aska	coscotoca ch'aska	illarimi ch'aska (rise ca. 4:00 a.m.)
$Songo_c$	ch'issin ch'aska		illarimi ch'aska
$Songo_d$		altopiña ch'aska	

I collected ethnoastronomical data recognizes a star which stands in the zenith at midnight. The zenith stars are also listed.

The name for the evening star is consistently given as ch'issin ch'aska ("evening star"); the morning star is referred to either as pachapacariq ch'aska ("dawn of the earth/time star"), illarimi ch'aska ("dawn star"), or locero (Sp. Venus). The terminological data will be analyzed, but for the moment we are still concerned with understanding the concept of twilight and its relation to the terms pacarin and ch'issin.

One way of analyzing the data in table 13 would be to attempt to define or identify the specific stars and planets referred to under each of the categories. However, this approach conflicts with the general point made thus far: that stars and planets are not given fixed, arbitrary names, but rather they are given names by which they are associated with units of space and time in relation to the sun. The importance of this concept will be understood better by shifting the focus of the argument for a moment to consider the relationship between twilight and the moon.

In González Holguín's dictionary of the Quechua language compiled and written in the sixteenth century, he gives the following names for the phases of the moon (González Holguín 1952):

a. chissi quilla. El tiempo en que alumbra hasta media noche, que es la creciente el primer quarto.	*a. chissi quilla.* The time in which the moon [= *quilla*] shines until midnight, which is the first quarter waxing.
b. paccar tuta, o tuta quilla. El segundo y tercero quarto de la luna que alumbra casi toda la noche.	*b. paccar tuta,* or *tuta quilla.* The second and third quarters of the moon which shine almost all night.
c. paccar quilla. Luna llena que resplandece en toda la noche.	*c. paccar quilla.* The full moon which shines all night.
d. paccarin quilla. El tiempo en que la luna alumbra a la mañana, que es la menguante el postrer quarto.	*d. paccarin quilla.* The time in which the moon shines in the morning, which is the waning last quarter.

When the terms given by González Holguín for the lunar phases are diagramed, as in figure 59, we find that the pacarin and ch'issin phases are equivalent to the two periods of twilight as described earlier. These data support the interpretation that pacarin and ch'issin refer to units of space/time rather than to specific celestial phenomena for neither term is used exclusively for a star, planet, or the moon. In addition, they provide us with a method for determining the length

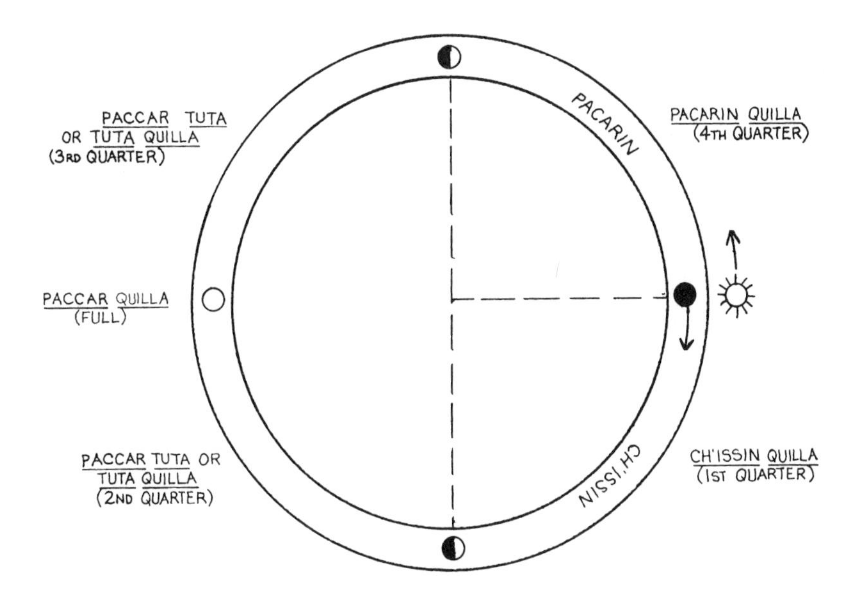

Fig. 59. Twilight and the Lunar Phases

of the two periods of twilight and of thus dispensing with the hour units in figure 56. The periods of twilight can be defined as one-half of the visible hemisphere or one-quarter of the total celestial sphere, the latter being equivalent to one-quarter of the lunar synodic cycle. This results in a quartering of the celestial sphere in accordance with the phases of the moon. However, as we see in González Holguín's lunar terms, the first and fourth lunar quarters are classified as separate units, whereas the second and third quarters are grouped together under one term (*paccar tuta* or *tuta quilla*). According to the terminology provided by González Holguín, which is in accordance with our earlier discussion of the wañu, cuscan, pura tripartite classification of the lunar synodic period (fig. 30), the "quarters" of the moon are actually thought of as "thirds":[2]

1st quarter	=	ch'issin	=	dusk (twilight)
2nd and 3rd quarters	=	paccar *or* tuta	=	night
4th quarters	=	pacarin	=	dawn (twilight)

This tripartite division of both the lunar synodic cycle and the night is consistent with the data in table 13, where we find a star of the dusk, one of the zenith at midnight, and one of the dawn. In addition, there is a similarity between the names for the zenith stars in table 13 and the names given by González Holguín for the combination of the second and third quarters of the moon.

Table 14. A Comparison of Lunar and Midnight Zenith Star Terminology

González Holguín	Table 13	
paccar tuta—"all night long"	cuscan tuta	ch'aska—"union of the two halves of the night star" or "midnight star"
paccar quilla—"moon which shines all night	coscotoca (tuta?)	
tuta quilla—"night moon"		

This association of the zenith star with the full moon is probably related to the easily observable fact that they will simultaneously rise at sunset, stand in the zenith at midnight, and set at sunrise. Regular observations of the changing shape and position of the moon, and of the relation between the moon and the twilight and zenith stars, allows the precise reckoning of time at night throughout the month. In addition, it provides the framework for integrating solar time (daytime) with the two units of time classified as twilight and the one unit of time classified as night (*tuta*). The use of midnight zenith stars in Quechua astronomy is in some respects similar to the use of *ziqpu* stars in Babylonian astronomy. That is, instead of actually observing the heliacal rise of star *A* one could instead observe star *B* (the *ziqpu* star), which stands in the zenith on the same morning that star *A* rises heliacally (Shaumberger-Gars 1952).

We now have a number of suggestions for how the lunar, solar, and stellar cycles may be integrated in the Quechua calendar. However, it is impossible to continue much beyond this point unless the above cycles and phenomena are incorporated into a larger framework. That is, the approach so far has allowed us to integrate a number of celestial phenomena and periodicities into a somewhat loosely correlated calendrical system, but it would be difficult to utilize such a system without relying upon continual astronomical observations and calculations. For the system to be manageable on a day-to-day basis, it must be placed within a more regular, easily observable framework. In effect, there must be some way whereby the system can operate without having to calculate all of the particulars. A kind of shorthand method for calculating astronomical time in Misminay will result from the unification of additional data on pachapacariq ch'aska with material presented earlier on the orientation of terrestrial and celestial space in accordance with the Milky Way.

The Four "Dawn of the Earth/Time Stars"

I have mentioned several times that the period from dusk to dawn is divided into three units, each of which is related to a star or planet and a unit of lunar time. In addition to this organization or system of three nightly stars, an informant in Misminay described an elaborate system of four stars, four pachapacariq ch'askas ("dawn of the earth/time stars"), each of which he related to one of the four terrestrial/celestial suyus (quarters). The informant defined the orientation of the appearance of the four stars as follows:[3]

a. *pachapacariq ch'aska of the eastern suyu*—appears around Calca or in the mountains to the north
b. *pachapacariq ch'aska of the southern suyu*—appears between Cuzco and Izcuchaca
c. *pachapacariq ch'aska of the western suyu*—appears around Apu K'otin, or a little more to the south (between Apu K'otin and Apu Quisqamoko)
d. *pachapacariq ch'aska of the northern suyu*—appears around Apu Veronica

These orientations are diagramed in figure 60.

In the drawing of the horizon positions of the four pachapacariq ch'askas, we find an intercardinal division and orientation of space similar to that discussed in chapter 2 in relation to the quartering of terrestrial and celestial space. This similarity will be discussed more fully after we have analyzed how the terrestrial orientations of the four pachapacariq ch'askas are integrated with the orientation of celestial units of space and time. The analysis first investigates the relationship between the four pachapacariq ch'askas and the three nightly stars described earlier in this chapter.

The point made repeatedly in connection with the evening, midnight, and morning stars is that they are related to the three periods of twilight → night → twilight. Since the four pachapacariq ch'askas are also twilight (i.e., "dawn") stars, how might an organization of four intercardinal twilight stars operate in conjunction with the division of the night sky, and the total period of the night, into three parts? This can best be answered by first changing the terminology for the concept of twilight from "twilight" to "heliacal." That is, in most ancient and primitive astronomical and calendrical systems, twilight is important as the time when one sees the heliacal rise and set of bright stars. From this point on, the *heliacal rise* is considered to be the first day on which a star is visible in the east before dawn; the *heliacal set* is taken as the first day on which a star sets in the west before dawn (see Aveni 1972:539). In both cases, the concern is with stellar observations made at *twilight*.

Chapter 2 established the Milky Way (= Mayu = River) as the principle phenomenon used in Misminay to orient celestial space. The questions now are how the Milky Way is used in the orientation of time and how this temporal orientation is related to the heliacal rise and/or set of the four pachapacariq ch'askas.

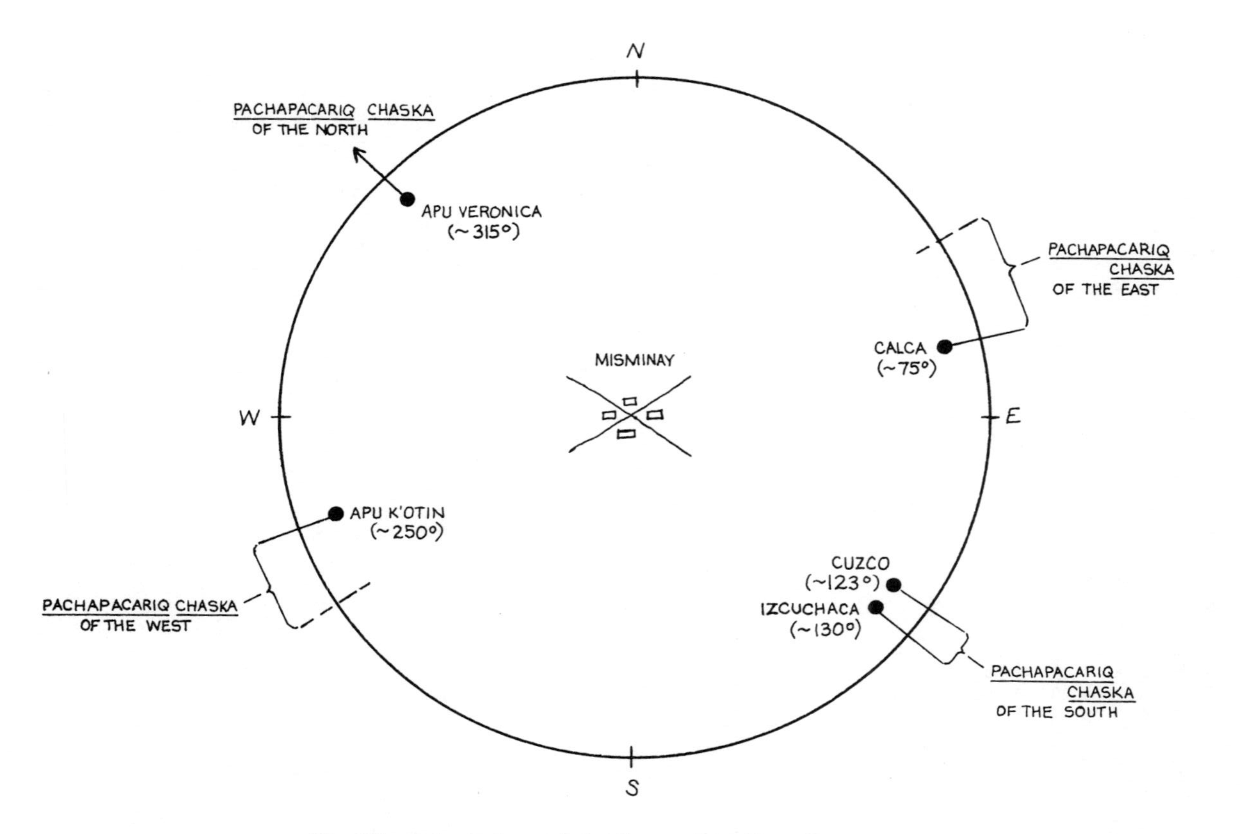

Fig. 60. Orientations of the Stars of the Four Quarters

Table 15 contains the heliacal rise and set dates of some of the principal stars located along the Milky Way (see fig. 61 for the locations of these stars).[4]

Table 15. Heliacal Rise and Set Dates of the "Quarters" of the Milky Way

Star	Heliacal Rise			Heliacal Set
α Crucis	Sept. 3			April 22
α Centauri	Oct. 9			May 21
Antares	Dec. 5	(Median =	(Median =	May 30
Altair	Jan. 28	*Dec. 17*)	*June 15*)	July 9
Deneb	Feb. 25			July 19
Capella	June 19			Nov. 28
Betelgeuse	June 18	(Median =	(Median =	Dec. 23
Procyon	June 29	*June 23*)	*Dec. 23*)	Jan. 17
ϵ Orionis	June 18			Dec. 22
Sirius	June 19			Jan. 8

The stars in the upper half of table 15 are those from α Crucis to the left in figure 61; the stars in the lower half are those to the right of α Crucis. Because α Crucis is at the center of the Milky Way, it will be considered separately. It is immediately apparent from the table that the median heliacal rise date of the stars in the upper half (December 17) and the median heliacal set date of the lower half (December 23) give a very close approximation of the December 21 solstice; the median heliacal-rise date of the lower half (June 23) and set date of the upper half (June 15) is very near the solstice of June 21. These diagonal relations come about because, as we have seen, the plane of the Milky Way is inclined with respect to the plane of rotation of the earth so that as the stars of one quarter rise, the stars of the *diagonal* quarter set (see fig. 19). These diagonal relationships result in a "quartering" of the ring of the Milky Way (fig. 61).

As Table 15 shows, the median heliacal rise or set date of each quarter of the Milky Way is related to the *time* of one of the two solstices, and this suggests that the four pachapacariq ch'askas mark the four places on the horizon where the "centers" (i.e., the median points) of each quarter of the Milky Way rise or set; thus, they ideally mark the *places* on the horizon which coincide with the *times* of the solstices. One easily observable phenomenon which leads us to this conclusion is that the median heliacal rise or set date of a quarter of the Milky Way will be the point in time which coincides with the midpoint of the line of that quarter of the Milky Way. Thus, I hypothesize that we can equate the mid-

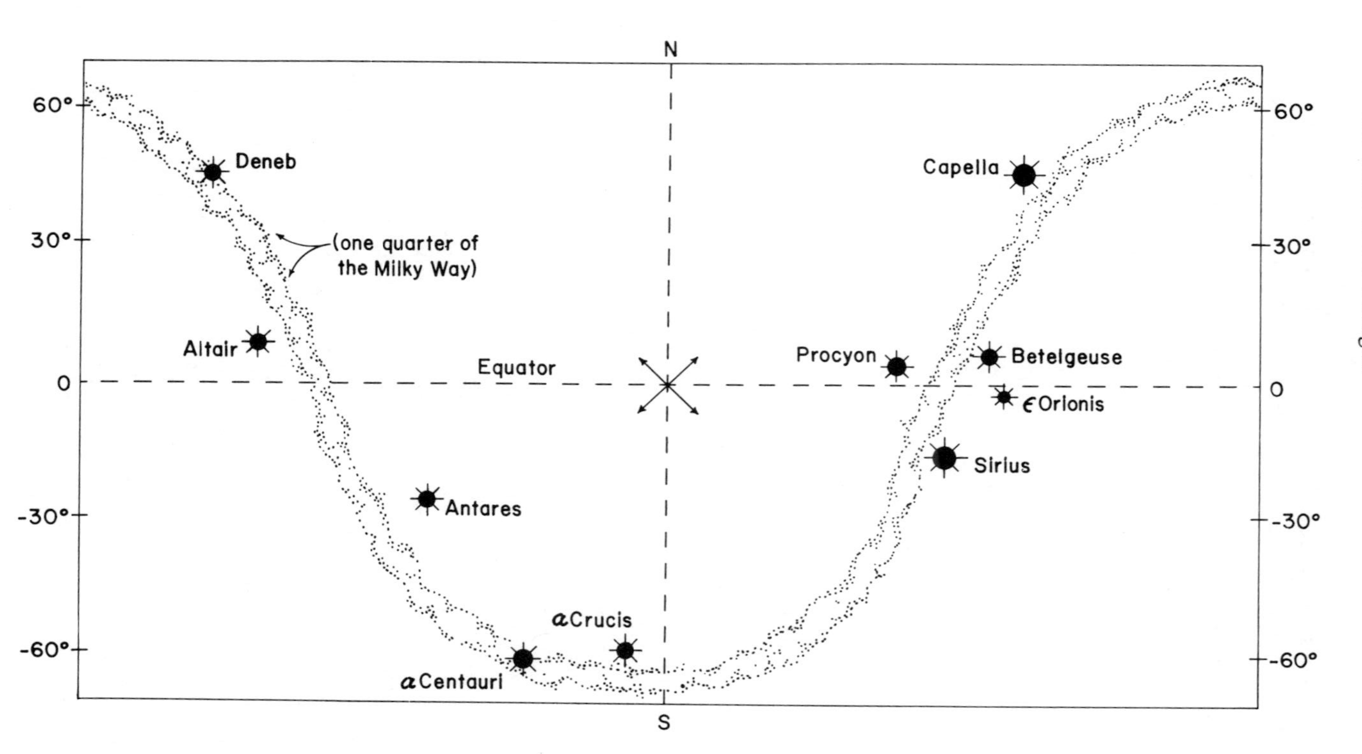

Fig. 61. The Quarters of the Milky Way

point in time with the midpoint in space. This can be done as follows.

In figure 19, we saw that the plane of the Milky Way is inclined with respect to the north-south axis of the plane of rotation of the earth. The inclination of the Milky Way from the north celestial pole is about 30°; the inclination from the south celestial pole is about 26°. This means that for an observer standing in Misminay, the Milky Way will not be seen within an area of 30° on *either side* of the north pole nor within 26° on *either side* of the south pole (see fig. 62; compare figs. 19 and 62). Therefore, to find the midpoints of the four quarters of the Milky Way, we subtract the 60° of space in the north and the 52° of space in the south within which the Milky Way will not be seen. Having done this, we arrive at the centers of the quarters of the Milky Way shown in figure 63.

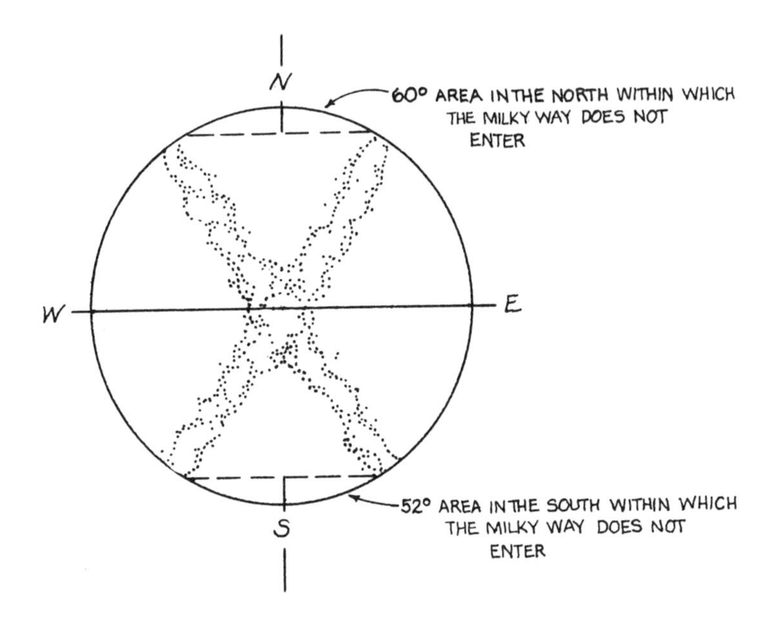

Fig. 62. The Limits of the Milky Way

Combining figure 60 with figure 63 produces the correspondences between the centers of the quarters of the Milky Way and the appearance points of the four pachapacariq ch'askas shown in figure 64. In all cases (except in the northwest) the appearance points of the pachapacariq ch'askas given by the informant in Misminay are very near the center points of the quarters of the Milky Way. On the basis of these correspondences, I hypothesize that the appearance

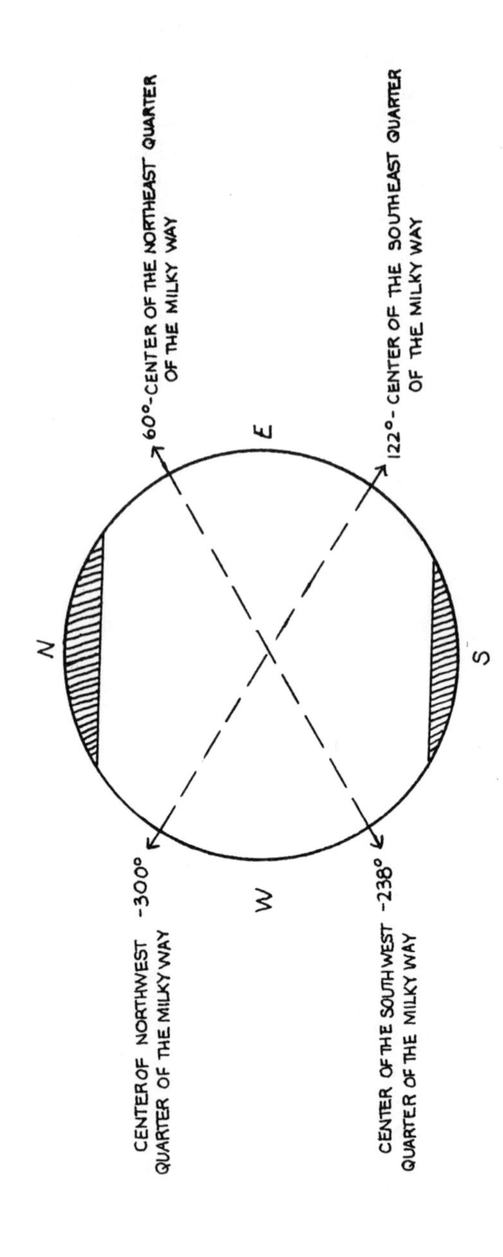

Fig. 63. Horizon Azimuths of the Centers of the Quarters of the Milky Way

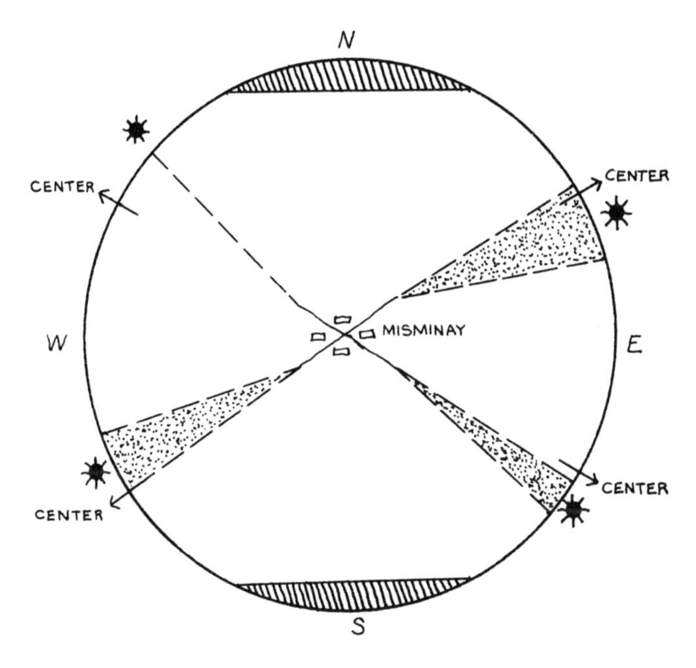

Fig. 64. A Second View of the Cosmology of Misminay

points given for the four pachapacariq ch'askas mark the four *places* on the horizon where the centers of the Milky Way rise or set *heliacally* at the times of the solstices. Thus, the times of the solstices can be calculated by observing the position of the Milky Way along the horizon at the two periods of twilight. For example, on the same morning that the center of the *southwestern* quarter of the Milky Way sets at dawn, the June solstice sun will be rising in the *northeast* above Urubamba. That evening, the June solstice sun will set in the *northwest* just as the center of the *southeastern* quarter of the Milky Way becomes visible in the mountains above Cuzco. In this way, the solar year can be easily reckoned by noting the changing relationships between the quarters of the Milky Way, the two periods of twilight, and the ring of the horizon.

At the beginning of this section, it was said that the above data would give us a possible schematic of the calendar system. To a certain extent this was accomplished by showing that the position of the Milky Way at dusk, midnight, and dawn will give a close reading of the time of day and the season of the year as they are related to the apparent annual movement of the sun and of the Milky Way. In addition, because the Milky Way forms a continuous line which moves steadily through the sky, it can also be used to calculate the sidereal and synodic cycles of the moon. Thus, the Milky Way provides a practical and easily understood orientational device for integrating various astronomical phenomena and cycles into an annual calendar. This is consistent with earlier discussions regarding the relationship between the seasons and the different orientations of the Milky Way.

The Planets

To this point, we have avoided what is probably the thorniest issue concerning the twilight stars. That is, are they stars at all? Or are they planets? Or can they be either stars or planets? On the one hand, we can say that if the twilight stars are associated with the solstices, as they appear to be from the analysis of the four pachapachariq ch'askas and in the data discussed earlier concerning the "mobile" star of Collca, then they may well refer to the planets (the planets, like the sun and the moon, move along the ecliptic). On the other hand, if the twilight stars are used for indicating the three regular divisions of the night, then they probably refer to stars since the cycles of the planets are so irregular. Perhaps a compromise is the case: the *three* stars of the three periods of the night are related to stars, whereas the *four* "dawn of the earth/time stars" are related to the planets.

In the consideration of the planets one further problem must be discussed because it is so crucial for the whole study of the modern Quechua and Incaic astronomical and calendrical systems. This is the problem of Venus. A number of characteristics of Venus make it an especially interesting celestial phenomenon. In the first place, its magnitude (at the brightest –4.3) is great enough to cast a shadow at night and to make it visible during the daytime. In Misminay, I was told by a number of informants that pachapacariq ch'aska can be seen during the daytime if one knows how to look properly. The "proper way" to look is by cupping the index finger into the base of the thumb to produce a tiny viewing slit. When the hand is held to the eye in this way, the amount of light entering the eye is greatly reduced and if one knows the exact position of Venus (and if it is in a crescent phase) the planet will be easily visible.

Another important characteristic of Venus is that it will never be seen *except* during the daytime at an altitude of more than about 47° above the horizon. This is a result of the fact that Venus orbits the sun between the earth and the sun; thus, Venus cannot be one of the midnight zenith stars mentioned in table 13, but it could be a *noon* zenith star. When Venus is seen near the sun during the daytime, it will always be located (unless it is in conjunction with the sun) within one of the two areas of space/time which precede or follow the sun (i.e., pacarin and ch'issin).

If, as mentioned above, the "star" which is visible during the daytime is called pachapacariq ch'aska, then why can we not positively conclude that pachapacariq ch'aska = Venus (since Venus is the only star or planet visible during the daytime)? In general, we can draw this conclusion, but with the stipulation that when Venus is not visible, it is "replaced" by a first magnitude star. On various occasions while in the field, informants referred to Venus, Sirius, and Canopus as the morning star (i.e., as pachapacariq ch'aska); in addition, one informant gave explicit *stellar* identifications for three of the four pachapacariq ch'askas (see n. 3).

It might be asked at this point why Venus cannot be replaced by another planet. The reason is that, every time a planet other than Venus was identified as the morning star (e.g., Jupiter and Saturn), it was called illarimi ch'aska ("dawn star"), *not* pachapacariq ch'aska ("dawn of the earth/time star"). On two different occasions, Jupiter and Saturn, both of which are *outer* planets, were pointed out to me as the midnight zenith star; again, Venus will never be referred to as the midnight zenith star because it is an inner planet. One informant gave a more elaborate definition of the term *illarimi* ("dawn"). He said that the term refers to a "star" which grows brighter as it passes all the way across the sky during the course of a night. I do not entirely understand this datum since it implies that when an outer planet is seen as the morning star, it will stand in the *west* at dawn, rather than in the east. Since Venus cannot pass across the sky at night, when the morning star is called illarimi ch'aska it must refer to an outer planet, but when the morning star is called pachapacariq ch'aska, it probably refers to an inner planet.

To summarize, the following equations seem reasonable:

a. pachapacariq ch'aska	=	Venus and/or a first magnitude star
b. illarimi ch'aska	=	An outer planet
c. zenith star	=	An outer planet and/or a first magnitude star which rises at sunset and sets at sunrise.
d. ch'issin ch'aska	=	Venus or an outer planet or a first magnitude star

9. Yana Phuyu: The Dark Cloud Animal Constellations

At the beginning of Juan Polo de Ondegardo's treatise on the "errors and superstitions of the Indians" written in 1571, he gives us one of our longest accounts of the constellations which were recognized by the Incas. Among the constellations are several animals and birds including llamas, a feline, and a serpent. By reference to other Spanish and indigenous chroniclers of Inca culture, the list of animal constellations can be expanded to include the tinamou, the condor, and the falcon.

In addition to the list of animal constellations, Polo gives us the following explicit statement concerning the relationship between celestial and terrestrial animals: "in general, [the Incas] believed that all the animals and birds on the earth had their likeness in the sky in whose responsibility was their procreation and augmentation" (Polo [1571] 1916:chap. 1; my translation).

The identification of the Incaic animal and bird constellations has eluded us for some time. In fact, this is not surprising because apparently their recognition even by the Incas was sometimes something of a problem. The chronicler Garcilaso de la Vega gives us the following confession concerning his early astronomical training.

They fancied they saw the figure of an ewe [llama] with the body complete suckling a lamb [uñallamacha], in some dark patches spread over what the astrologers call the Milky Way. They tried to point it out to me saying: "Don't you see the head of the ewe?" "There is the lamb's head sucking"; "There are their bodies and their legs." But I could see nothing but the spots, which must have been for want of imagination on my part (Garcilaso [1609] 1966:119 [bk. 2, chap. 23]).

What was apparently the same llama, and its suckling baby, is also described in the chronicle of Francisco de Avila (1608) for the central Andean community of Huarochirí. The llama, says Avila, was "blacker than the night sky" (Avila [1608] 1966:chap. 29).

From the testimony of Garcilaso and Avila alone it should be clear that in order to identify the animal constellations of the Incas we should look first to the "dark spots." However, as mentioned in the Introduction, the literature on Incaic ethnoastronomy is full of attempts to explain away, or to dismiss entirely without comment, the testimony on the dark cloud animal constellations in the Milky Way (e.g., Lehmann-Nitsche 1928:36).

Constellations of this type have only rarely been reported in the ethnoastronomical literature, and one of the goals here is to present as complete a description as possible in light of the data collected in Misminay in order to arrive at an understanding of the general principles of Quechua astronomy and cosmology incorporated in this one category of celestial phenomena.

Celestial Locations and Periodicities of the Yana Phuyu

As described earlier, the "dark cloud" (yana phuyu) constellations are located in the southern portion of the Milky Way where one sees the densest clustering of stars and the greatest surface brightness and where, therefore, the fixed clouds of interstellar dust which cut through the center of the Milky Way (the dark cloud constellations) appear in sharp contrast. The dark cloud constellations identified by informants in Misminay and elsewhere in the Cuzco area are as follows, listed in the order in which they rise along the south-eastern horizon (see fig. 65):

Mach'ácuay – Serpent
Hanp'átu – Toad
Yutu – Tinamou
Llama – Llama
Uñallamacha – Baby Llama (/or llama's umbilicus/or serpent)
Atoq – Fox
Yutu – Tinamou

From the head of the Serpent in the west to the tail of the Tinamou in the east, the dark cloud constellations stretch in a line through about 150° of celestial space, straight along the central course of the Milky Way. Figure 65 shows all of the dark cloud constellations in the sky at the same time. The view is toward the south from a hypothetical location in the southern Andes near the community of Misminay (i.e., Misminay's latitude is –13°30′; thus, as shown in the drawing, the unmarked south celestial pole stands 13°30′ above the southern horizon).

As the sky appears to revolve at night throughout the course of a year, it will

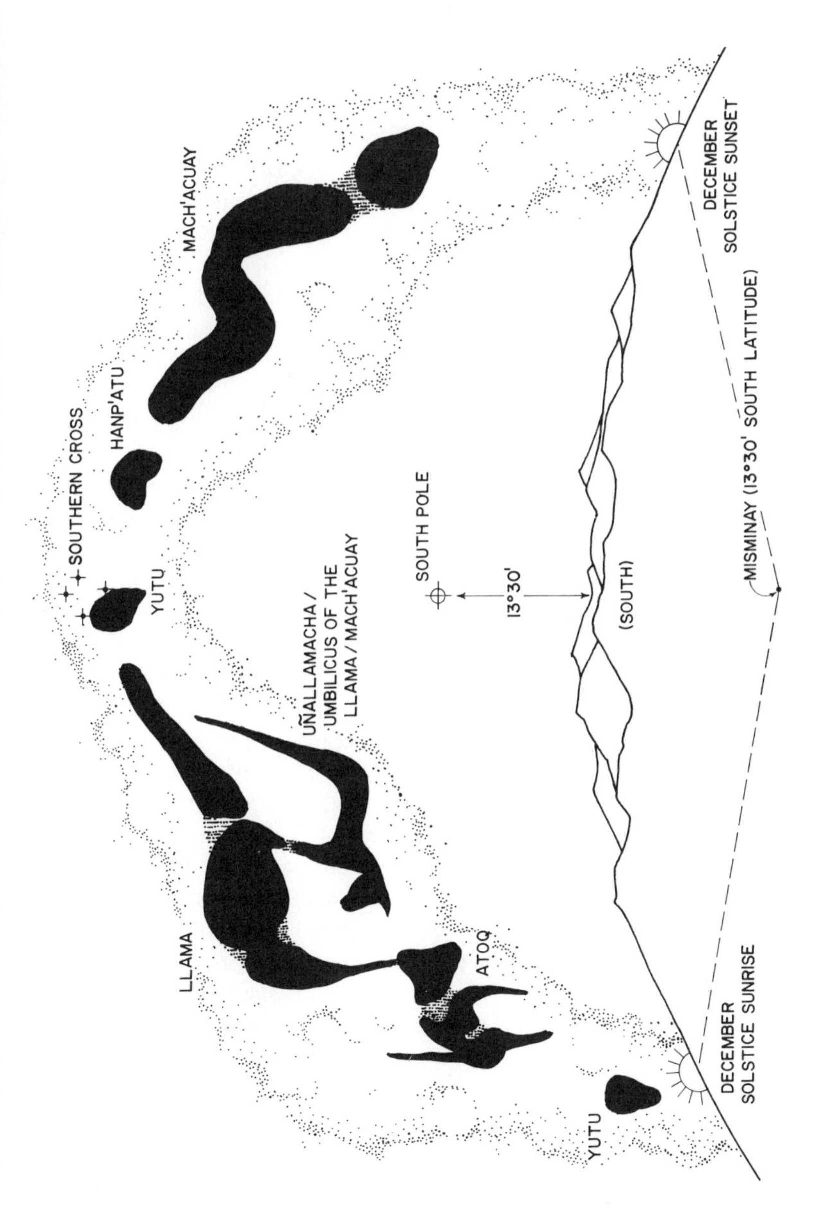

Fig. 65. The *Yana Phuyu* Viewed from Misminay

be unusual to actually see the entire line of dark cloud constellations in the sky on the same night. The time of greatest visibility occurs when the center of the line, the area around the Southern Cross and the Yutu, stands along the north-south meridian at midnight; this occurs around the twenty-third of March, the day of the autumnal equinox in the southern hemisphere. If we rotate the sky so that the entire line of dark cloud constellations is underground at midnight, which will therefore be the night on which the *fewest* dark clouds will be seen, we arrive at a date of September 26, very near the vernal equinox.

In addition to this temporal relationship to the two equinoxes, we see in figure 65 that there is an orientational relationship to the December solstice; that is, the most extreme *northerly* rising and setting points of the dark clouds coincide with the most *southerly* rising and setting points of the sun. Thus, at midnight on the night of the March equinox, the line of dark cloud constellations stretches in an arc through the southern skies from the rising point in the southeast to the setting point in the southwest of the December solstice sun.

Another important observational and temporal characteristic of the dark cloud constellations is their relationship to the rainy season, which begins around October and ends in early April, with the heaviest period of rains lasting from December through February. In discussing one of the dark cloud constellations (the Serpent), informants in Misminay said that the dark cloud Serpent is seen in the sky at night during the rainy season but during the dry season (May-July) it is below ground at night. Besides being related to the rainy season, the dark cloud constellations are associated in another direct way with water, for they are located along the central course of the Milky Way.

As discussed in more detail in chapter 2, the celestial River, the Milky Way, is believed to carry into the sky the actual water which flows through the Vilcanota River. As the Vilcanota River flows from the southeast to the northwest, it carries terrestrial water to the edge of the earth. The water then flows into the mar, the cosmic sea, which completely encircles the earth. As the Milky Way revolves around the earth, it dips into the cosmic sea in the west, takes in the terrestrial water, passes underground, and rises again in the east. The Milky Way moves slowly through the sky above the earth, depositing water throughout the celestial sphere. The water then returns to the earth in the form of rain, where, in its continuous cosmic cycle, it again flows along the tributaries which feed into the Vilcanota River. In this way, the celestial and terrestrial rivers act in concert to continuously recycle water, the source of fertilization, throughout the universe.

It is clear from this description of the cycling of the Milky Way that in order to understand the full significance of the dark cloud animal constellations within it, we must consider the three elements, or parts of the universe, with which they are therefore associated: sky, water, and also earth (since they are subterranean for at least half of every day). A number of additional factors must be considered, such as the specific types of animals represented, their biological cycles and behavior patterns, and the question of the correlation between astronomical and biological cycles as suggested in the quotation from Polo de Ondegardo.

Of the three factors mentioned above—sky, water, and earth—water occupies a position of mediation, for it is the element cycled through the other two. It is therefore essential to study the Dark Cloud constellations by analyzing the connections between sky and water and earth and water.

Sky and Water

Alexander von Humboldt, in his monumental studies of the natural sciences in South America, made the following observation: "I have endeavoured to describe the approach of the rainy season, and the signs by which it is announced .... The dark spot in the constellation of the Southern Cross becomes indistinct in proportion as the transparency of the atmosphere decreases and this change announces the approach of rain" (von Humboldt [1850] 1975:138-139).

In Misminay, I also was told that the dark cloud constellations are observed in the prediction of rain. Although informants in Misminay did not describe the specific methods used, nor the times when the observations are made, explicit testimony from elsewhere in the southern Andes indicates that such predictions are made during the month of August at the beginning of the planting season and at the time of transition from the dry season to the rainy season. Padre Jorge Lira has recorded the following meteorological and crop predictions in use today in communities around Cuzco: "If the stars of the sky appear bright and beautiful, everything will be good, materially and spiritually. If, in the Milky Way, there is an *accentuation of the dark areas*, or the 'sacks of carbon,' it will be a year of pestilence and death" (Lira 1946:18-19; my translation and emphasis).

Combining the accounts of von Humboldt and Lira leads to the conclusion that, if the *obscuring* of the dark clouds indicates the approach of rain, their *accentuation*, as described by Padre Lira, indicates the absence of rain. Therefore, we find a curious kind of inversion in the relationship between water and the dark cloud constellations; their *appearance* in the night sky is associated with the period of the rainy season, but their gradual fading out, or "*disappearance*," as a result of increasing atmospheric moisture announces the actual approach of rain. Because the dark cloud constellations are located within the celestial River, which spreads water from the cosmic sea throughout the celestial sphere, the weather predictions described by Padre Lira reflect a coherent and logical explanation of the operation, and interrelation, of certain natural phenomena. But beyond the "convenience" of a consistent explanation of the natural universe, the data and processes outlined above provide a system of prognostication, an essential element in the survival of communities whose livelihood depends on the success of the crops, which in turn depends on the amount of rainfall.

These observations on the role of the dark cloud constellations in relation to agricultural prognostication must be seen in connection with the methods discussed in previous chapters. We have found that Collca (the Pleiades) is observed both near the time of the June solstice and at the beginning of August for the purposes of determining not only how good the harvest will be but also for knowing precisely when to plant. The Collca observations are in turn related to the waxing (planting) moon and to the rise of the sun between two points along the eastern horizon (i.e., Calca and Chinchero) as it moves southward. In short, the observations of the relative brightness and obscurity of the dark cloud constellations are only one part of a complex system of observing a number of natural signs which provide the runa of Misminay with a broad range of indicators of continuity and change in the physical universe.

Earth and Water (*Pachatira*)

The water within the celestial River has a terrestrial origin, so it is not surprising to find that the animals of the Milky Way also originate from the earth. According to one informant in Misminay, the yana phuyu ("dark clouds") are actual pieces of earth taken up into the sky by the Milky Way. The informant was uncertain as to whether the animals are taken up under the earth, during the subterranean passage of the Milky Way, or enter it from the mountain tops, where, he said, there are a lot of wild animals.

The terrestrial origin of these celestial animals is further indicated by the fact that even though they are located in the sky, they are classified as *pachatira (pachatierra),* a name which combines the Quechua and Spanish words for "earth." The name Pachatira was obtained in a situation which throws additional light on the symbolic significance of the term as used in relation to celestial phenomena. Recall the long conversation I had with a group of men and womed in Misminay, when I asked about the sexual association of various astronomical bodies. It was generally agreed that single stars, as well as the star-to-star constellations, are masculine. When asked about the dark cloud constellations, one man answered immediately that they are female. Later, however, he pointedly returned to the question and said that he had been wrong earlier in calling them female; they are, he said, pachatira. Thus, though the dark cloud animals may be thought of as more female than male in opposition to the stars which are male, they are actually neither—they are pachatira.

Pachatira is an important concept in Quechua cosmological thought. In the community of Kuyo Grande, Casaverde Rojas found that Pachatierra is classified as female and is considered to be the malevolent twin sister of Pachamama, "Earth Mother" (Casaverde Rojas 1970:150). Oscar Núñez del Prado (1973:36) gives the following description of the malevolent nature of Pachatierra as found in Kuyu Chico: "She is wicked and eats the hearts of men, who then die spitting blood. She is generally found by cliffs and precipices, and her preferred victims are children or adults who stay asleep in bad weather."

The femaleness of Pachatira, and the relationship of Pachatira to the Earth Mother (Pachamama), is also found in the area of Ocongate, where three feminine forms of earth (*pacha*) combine to express the total concept of Pachamama; the three are Pacha Tierra, Pacha Ñusta, and Pacha Virgen (Gow and Condori 1976:6).

In the community of Sonqo, when blowing a coca *k'intu*, one often calls on "Mother Earth," which includes Santa Tira, Pachamama, and Pachatira Mama (Wagner 1976:200). No clear-cut distinction is made in Sonqo between Pachamama and Pachatierra; they are both female and both related to *hallp'a* and *pampa* ("soil, ground"; Catherine Allen, personal communication, 1978).

From these accounts, we find that Pachatira refers primarily to the earth and to its powers of fecundity. It is also often associated with a concept of femaleness in relation to Pachamama, but the latter does not appear to be a necessary nor invariable characteristic (see J. V. Núñez del Prado 1970:75-76). This slight ambiguity in the sexual classification of Pachatira is well illustrated by the man in Misminay who first called the dark cloud constellations female, but then insisted that they are more properly classified as pachatira.

That *pachatira* refers to a concept of earthly or subterranean fecundity is important to our discussion of celestial animals which originate from, and are actually composed of, the earth. We find in the Andes a general belief in the subterranean origin of all animals (see Aranguren Paz 1975:108 and Duviols 1976:283). As stated by an informant of the Gows' in the area of Ocongate,

It was a very long time before alpacas existed. When it first dawned, they were hidden under the earth where there are springs. Then, when the sun rose again, all the animals came out of a spring. For this reason, we make an offering to a spring and the lakes at the foot of Ausangate. If there had been no subterranean spring, we would not have had animals. The spring and the lakes are the owners of the animals (Gow and Gow 1975:142; my translation).

With this quotation, we can begin to understand the relation not only between animals and the earth (pachatira), but also between animals and earthly, or subterranean, *water*. Since the animals of the earth, those actually used by humans for food, clothing, and transport, originate from subterranean springs, it is not surprising to find that the animals which inhabit the waters of the celestial River are also related to the concept of the earth as a fecund force (i.e., to pachatira). As noted earlier in the description of the diurnal rotation of the Milky Way, the celestial River passes beneath the earth after first entering the cosmic sea in the west. We can well imagine the tremendous mixing and crossing of subterranean water, earth, and animals which occurs as the Milky Way passes beneath the earth and how, therefore, the animals in the sky are intimately connected with the animals of the earth.

Rainbows have been found to mediate between subterranean water and "human water," urine (chap. 4). This indicates that a more general conceptual relation might exist between urine and subterranean water. We are as yet far from understanding the symbolism of urine in the Andes, but the following indicates that urinary symbolism is an important field of study. In Sonqo, the *q'oncha* ("fireplace"), which is the domain of women, is made of a mixture of female urine and ashes (Catherine Allen, personal communication, 1978). In Misminay I once watched as a large black bull was forced to drink male urine. The bull was hobbled, tied to a stake, and brought to the ground by five or six men. He was then held on the ground while two or three of the men in turn urinated into a small bowl and force-fed the urine to the bull. I was told that this is done when a bull needs to be fattened up or made braver (*bravo*) for breeding. It may also be significant that a *black* bull was forced to drink urine (which is equivalent to human/subterranean water). About the relation between black bulls and water, Alejandro Ortiz Rescaniere (1973:29 n. 1) says that "the black bull in the Andes is considered to be an ill-omened and diabolical animal . . . , the black bull frequently represents rain which is harmful to agriculture" (my translation). These data suggest that pachatira may be considered not only as the power of the fecundity within the earth, but also, by extension, as the internal power of fecundity of human and animal bodies.

With these observations on the general nature of the category of dark cloud constellations as a background, we can now examine the specific characteristics of each of the terrestrial and celestial animals in order to determine what relationship exists between the behavior of the animals and the behavior of the constellations. Our investigation is prompted by the statement of Polo de Ondegardo to the effect that the Incas believed that the animal constellations were responsible for the "procreation and augmentation" of their animal counterparts on the earth. We begin with the Serpent, at the head of the line of dark cloud constellations, and proceed eastward along the line of constellations as they rise after the Serpent. We will consider all of the animals and birds shown in figure 65, except for the Tinamou located on the left (east) side of the drawing. Only one informant placed a second Tinamou in this position and its precise location must be determined in future fieldwork.

Mach'ácuay ("serpent"; *SC*# 41)

Snakes are relatively rare among southern Andean fauna, especially in light of the fact that one of *the* herpetaria of the world, the Upper Amazonian rain forest, lies just to the east and north. The only naturally occurring species of snake above twelve thousand feet in the Andes is the mildly poisonous Colubrid, *Tachymenis peruviana. T. peruviana* ranges in size up to about one-half a meter, and its coloring is a yellowish or pale brown with dark spots and longitudinal streaks running along the upper part of its body. An oblique dark streak runs from the eye to the angle of the mouth. A pair of grooved fangs are positioned below the back of the eye (Boulenger 1961:117-119). *T. peruviana* gives birth to live young and its birthing period is between September and October (Fitch 1970:156). Its altitudinal range is between six thousand and fifteen thousand feet.

The term *mach'ácuay* is commonly found in the literature on the reptilian fauna of the Department of Cuzco, and I suggest that it refers primarily to the indigenous snake *T. peruviana*. The following two reptiles have also been described for the Cuzco area and may be considered as varieties of the *mach'ácuay* group: *yana-muroq* ("dark spotted"), a small green and white snake, and *oje-muroq* ("grayish spotted"), a small brown-to-dark-brown snake. The more impressive South American reptiles, such as the anaconda and the boas, do not occur naturally in the area. However, Garcilaso de la Vega (bk. 5, chap. 10) tells us that the inhabitants of the Inca empire who lived in the jungle to the east of Cuzco brought huge reptiles (*amarus*) to the Inca as tribute. In addition, the present distribution of the medium-sized *Constrictor constrictor ortonii* along the Upper Marañón and of the *Boa hortulana hortulana* of the Upper Madre de Dios east of the Department of Cuzco (see Schmidt and Walker 1943a: 280 and 1943b:305) are good indications that the large Amazonian reptiles are known today by the inhabitants of the high Andes.

In Misminay, it is common to travel far down the Urubamba Valley for work and to consult the shamans of the lowlands for purposes of divination. Daniel Gade (1972) has produced ample documentation to support the fact that the Urubamba River has for quite some time served as a major artery for travel and trading between the Indians of the highlands and lowlands in southern Peru. This pattern of lowland travel was common among pre-Columbian Andean populations, and it is entirely feasible to suppose that the medium-to-large reptiles of the Upper Amazon have been a part of the faunal knowledge of Andean peoples for some time.

In addition to the Serpent constellation called Mach'ácuay, another apparently serpentine constellation, Sullu-ullucu (*SC#* 44), was mentioned by an informant immediately after a discussion of Mach'ắcuay. The word *sulluullucu* may be related to the name Surucucu, the deadly Bushmaster (*Lachesis mutus*) of the tropical forest (see Ditmars 1937:137 and Tastevin 1925:172). If this derivation or relationship of the Quechua term is correct, it is all the more impressive since the constellation Sullullucu was mentioned in the community of Sonqo, at an altitude of about 14,500 feet. A further suggestion that large Amazonian serpents are known in the Andes today is found in the Quechua dictionary of Padre Jorge Lira. Lira gives the term *mach'ắcuay* for "serpent," but he also gives a name for the boa, *A'ti mach'akkway*. Lira relates the name A'ti mach'akkway to *amaru*, the monstrous serpent of Andean mythology and also, as mentioned, the large serpents given as tribute to the Incas. In a document written by Cristóbal de Albornoz (see Duviols 1967:23), *mach'ắcuay* and *amaru* were together associated with the Inca ruling class: "There is another type of *guaca* [holy object] which is a certain type of snake having different forms. These . . . [the Incas] adore and serve, and the principal Incas use them as their names. They are called mach'ắcuay and amaru" (my translation).

The term *amaru* is important in Quechua meteorology because it is used for rainbows, which are believed to be giant serpents (chap. 4). The body of the rainbow Serpent rises up out of one spring, arches through the sky, and buries the opposite end of its body in another spring. Amarus are thought of as double headed; one head is buried in each spring.

These data relate the dark cloud constellation serpent, Mach'ắcuay, and the rainbow serpent, Amaru; this is a relation of dark and multicolored reptiles. Precisely the same relationship or opposition of coloring is found in the ethnoastronomy of tropical forest reptiles. As described in P. C. Tastevin's study of the Amazonian legend of Bốyusû (= Surucucu, the Bushmaster), "the celestial Bốyusû appears during the day in the form of the rainbow, and at night in the form of a dark spot" (Tastevin 1925:183; my translation). Tastevin further shows that the dark-spot serpent is a dark cloud in the Milky Way which winds itself round the constellation of Scorpio. Scorpio is important in the Amazonian calendar because, when it stands in the zenith in November, it signals the beginning of the rainy season (Tastevin 1925:173).

In the mythology of the Guaraní and Guayakí Indians of Paraguay, rainbows are thought to have two aspects, one multicolored and the other black. In addition, the giant Rainbow/Serpent of the Guaraní is related to a celestial "nebula" (see Cadogan 1973:98-99) which, we would suspect, is a *dark* nebula. Celestial serpents in the Amazon are therefore related to rainbows (water), dark clouds in the Milky Way, and to the rainy season. This is the same complex of associations found in Andean astronomy:

mach'ắcuay—dark cloud in the Milky Way observed at the beginning of the rainy season

amaru – the double-headed, multicolored rainbow serpent

The emergence of "meteorological" serpents (amarus) from the ground immediately following a rain shower, and their reentry as the atmosphere becomes less moist, is an important clue to understanding the relation between terrestrial and celestial reptiles in the Andes. The amaru which rises out of a spring after rain, exhibits a climatological behavior pattern similar to terrestrial serpents, which, at the *end* of the cold/dry season and at the *beginning* of the warm/rainy season, emerge from subterranean hibernation. The Andean dry/cold season (May-July) is a period of reduced activity not only among reptilian fauna but also among the fauna on which reptiles prey. Therefore, terrestrial reptiles in the Andes are variably active and inactive in direct relation to pronounced alternations between dry/cold and warm/rainy seasonal changes (see Schoener 1977: 115-116). Because meteorological serpents (rainbows/amarus) appear only during the rainy part of the year, the exhibit a seasonal activity cycle similar to that of terrestrial reptiles.

The principal identification of the dark cloud Serpent (Mach'ácuay, see figs. 33 and 65) is a large zig-zag-like streak of interstellar dust which stretches from a point near the Southern Cross to Adhara (in the Western constellation of Canis Majoris). The head of the Serpent precedes the tail in rising; thus, the movement of the constellation through the night sky can be likened to terrestrial serpents and rainbow serpents, which rise out of the earth headfirst and reenter headfirst. Because the dark cloud Serpent stretches over such a large celestial area, one part of its body must be chosen as a point for analysis of the cycles of visibility and invisibility of the celestial Serpent. For this purpose a good choice is the head of the Serpent, for the head is crucial in determining the time and place of the emergence and reentry of Mach'ácuay. The heliacal rise of the head of the Serpent occurs during the first week of *August;* its heliacal set occurs during the first week of *February.* The most intense period of rain in the southern Andes occurs between the months of December and February, and the planting period begins at the beginning of the change from dry to rainy in August. Therefore, the periodicity of the celestial Serpent's rising out of the earth and reentering it during the night brackets the rainy season. In effect, the celestial Serpent, like meteorological serpents, emerges from the earth with the warm/rainy season and reenters the earth at the beginning of the dry/cold season. In addition, the principal serpent of the Cuzco area above twelve thousand feet, *T. peruviana,* gives birth to its young from September to October, just after the onset of the warm/rainy season. This analysis suggests that Polo de Ondegardo's statement concerning the responsibility of celestial animals for their terrestrial counterparts refers to the easily observable, and cosmologically important, correspondence between the periodicity of the presence and absence of terrestrial, celestial, and meteorological reptilian fauna in the universe.

Hanp'átu ("toad"; *SC*# 37)

That the celestial Toad appears to "pursue" the Serpent across the sky is ironic in light of the fact that snakes are the greatest predators of toads and frogs. Certain toads are occasionally known to get the better part in combat with snakes, but this is by no means the usual outcome. In Quechua a distinction is made, as in the classification of Amphibia here, between the more aquatic frogs (*ococo, k'ayra,* and *ch'eqlla*) and the terrestrial toads (*hanp'átu*). Toads rather than frogs are the primary concern, since every reference I collected or have encountered in the literature regarding a celestial Amphibian has been in relation to hanp'átu, the toad (see Roca 1966:43; Cobo 1964, vol. 1:352; and González Holguín 1952).

The principal Andean toad is *Bufo spinulosus. B. spinulosus* tolerates dryness and altitude very well and breeds principally at the onset of the rainy season in permanent bodies of water. The range of *B. spinulosus* in South America extends along the cordilleras southward to –43° latitude; the altitudinal range to which they have adapted is between one thousand and five thousand meters (Cei 1972:83).

In a study of the behavior of assorted fauna on the plain of Anta, southeast of Misminay (see map 1), Demetrio Roca (1966:45) says that in addition to the name *hanp'átu*, toads are also referred to by the following names:

Pachakuti – "turning of the earth"
Saqra – "devil"
Pachawawa – "earth child"
Jacinto – "hyacinth"

Toads are called "devils" (*saqra*) because they were created by the devil, because they foretell bad luck when seen (Cobo 1964, vol. 1:353), and because they are used in the malevolent practices of witches (Roca 1966:45). The two terms *pachawawa* and *pachakuti* are important for our study because they refer to the common habit of toads to burrow within the earth during the dry/cold season and, like mach'ácuay, to reemerge with the warm/rainy season (see Noble 1931:421 and Grzimek et al. 1974:360-367). It is also important to note that Amphibia are most active at night, when the humidity is much greater than during the day. Therefore, toads are the "children of the earth" (*pachawawa*) in that they hibernate within, and later emerge from, the earth. This cyclical entry and reemergence, coinciding with the cycling of the dry/cold and warm/rainy seasons, is a behavioral pattern well described by the name *pachakuti* ("turning of the earth").

From fieldwork and observations of the behavior of toads on the plain of

Anta, Roca (1966:42) gives this description of this cycle of subterranean hibernation: "The earth is alive during the month of August, being intensely animated by the toad or *pachakuti*, which emerges from the interior of the earth in great numbers. It is noted in the plain [of Anta] that beginning in the month of May, wide deep cracks appear in the earth through which the toads return to the womb of the earth, reappearing in the month of August" (my translation). After the initial emergence of the toads in August, their behavior (mating, croaking, and so on) is observed closely for divinatory purposes: "If in the months of September and October they croak day and night in great numbers, it is an augury that there will be much rain and, as a consequence, the crops will be abundant; but, if during these months they croak only a little and softly, it is a sign that it will not rain and that the frosts will be strong" (Roca 1966:58-59; my translation).

In addition to a divinatory connection between toads, weather, and the crops, toads are also related to agriculture because at the beginning of the planting season in August, toads encountered in great numbers indicate that the year's crop will be abundant; if only a few small ones are encountered, the crop will be small (Roca 1966:59).

A summary of the cyclical behavior of toads yields the following calendar of activities in relation to agriculture and the seasonal cycle (fig. 66). In this "calendar of toads and agriculture" the cycles of agriculture and Amphibian behavior closely coincide. The points of transition occur along the temporal boundaries defined by the cycling of the cold/dry season and the warm/rainy season. As in the discussion of Mach'ácuay, a correlation is found in the relationship between the seasonal and the terrestrial and celestial toads.

The dark cloud constellation of Hanp'átu is a small patch of interstellar dust in the Milky Way which moves between the tail of Mach'ácuay and the Southern Cross (see fig. 65). Consulting a celestial globe, we find that during the first days of the month of October (i.e., at the time of the mating period of terrestrial toads), Hanp'átu rises about one and one-half hours before the sun; thereafter, Hanp'átu rises progressively earlier than the sun each morning. In effect, the celestial toad rises into the sky in the early morning just after terrestrial toads have emerged from their long period of subterranean hibernation and just at the time of their most intense croaking and mating period.

Yutu ("tinamou"; *SC*# 47)

Near the center of the line of dark cloud constellations (see fig. 65), at the point where the Milky Way flows nearest to the south celestial pole, sits the Tinamou (*Yutu*). The Yutu of Quechua astronomy is equivalent to the Western constellation of the Coalsack, one of the few dark spots recognized, and named,

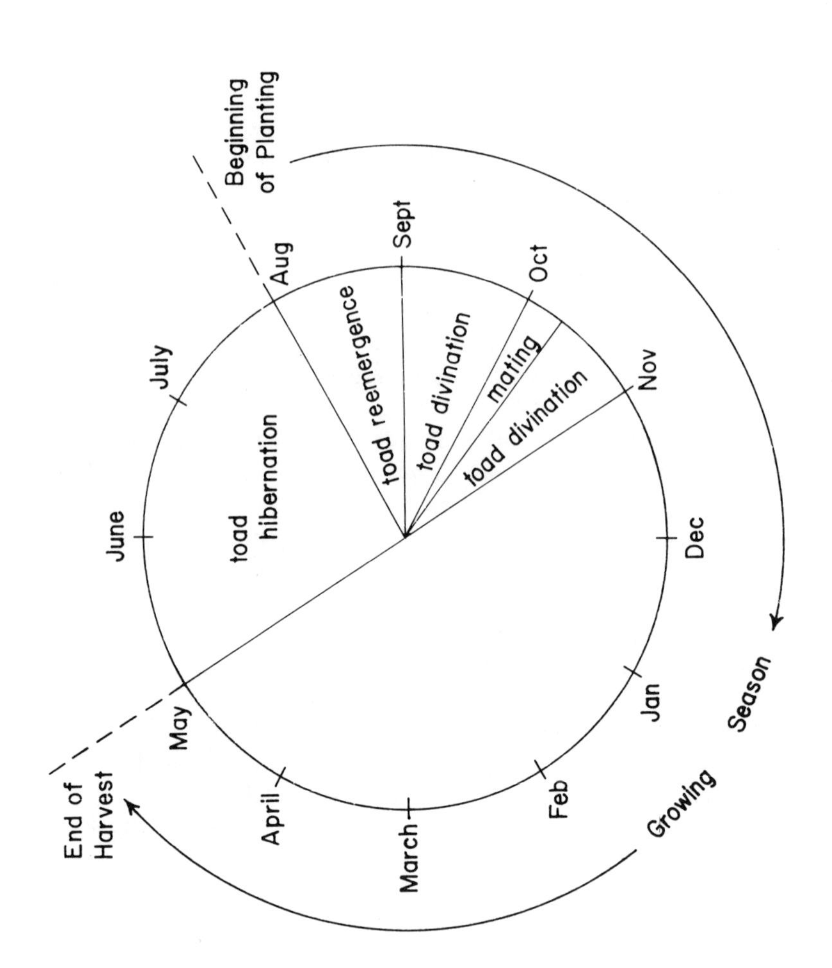

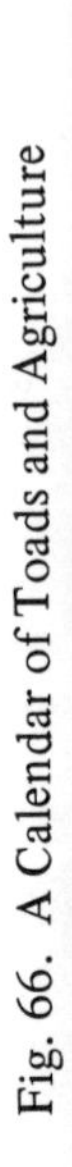
Fig. 66. A Calendar of Toads and Agriculture

in our own system of astronomy.

The partridge-like birds called Yutu (Quechua) and Tinamou (Carib), resemble game birds with their short legs, compact bodies, and small heads with slender necks. There are some nine genera and forty-three to forty-five species of tinamous distributed throughout South America and northward to the Tropic of Cancer (Grzimek et al. 1972:82 and Lancaster 1964).

The range of vertical habitat of tinamous extends from the tropical rain forests up to the high, cold punaland of the Andes (de Schauensee 1970:309, Roe and Rees 1979:475-476, and Traylor 1952). Tinamous eat mostly seeds and fruits, although they are occasionally known to swallow small animals whole. D. A. Lancaster (1964:part a, p. 171), for instance, observed tinamous eating frogs and lizards on several occasions. This observation is especially interesting in the present context, for the dark cloud constellation of the Tinamou (Yutu) "pursues" the celestial Toad through the sky. In a somewhat less antagonistic characterization, one informant described to me the nightly race between the Toad and the Tinamou. The Toad, said the young woman, always wins the race and therefore, her husband likes to characterize himself as a toad in opposition to others who are tinamous (see Urton 1978a).

The characterization of the Yutu as a slow animal in relation to other celestial animals is based perhaps not only on these nightly celestial races, but also on the terrestrial behavior of tinamous. Tinamous are notoriously slow, "stupid" birds (Cobo 1964, vol. 1:321). Not only are they disinclined to flight, but when flushed, they fly low and poorly. Long ago, W. H. Hudson made the following observations on the flight of tinamous.

> The Tinamou starts forward with such amazing energy, until this is expended and the moment of gliding comes, that the flight is just as ungovernable to the bird as the motion of a brakeless engine, rushing along at full speed, would be to the driver . . . . In the course of a short ride of ten miles, I have seen some of these Tinamous dash themselves to death against a fence close to the path, the height of which they evidently misjudged. I have also seen a bird fly blindly against the wall of a house, killing itself (quoted in Knowlton 1909:78-79).

In fact, the most common reaction of the Boucard tinamou when startled is not to fly but rather to do nothing, to simply "freeze" (Lancaster 1964:part a, p. 171). Barring escape by freezing or flying, a tinamou, hard pressed in open country, will often crawl into a hole dug by another animal.

Aside from the above "un-bird-like" features, which no doubt make tinamous a noticeable part of the bird population in Andean communities, a number of additional characteristics may contribute to its celestial projection in the form of a dark cloud constellation. First, tinamous are distinctive in their solitary, unsociable nature. They are rarely found in coveys, either with members of their own or other species. This unsociable or solitary behavior is extended to breeding and incubation habits. At the beginning of the mating season, males begin calling in the early mornings and late evenings. This pattern increases until the height of the breeding period, after which time calling begins to subside. During the breeding period, males attract a number of females to their nests by this persistent calling. The females all lay their eggs in the male's nest and depart, wooed to another male's nest by his flutelike whistle. After several (two to five) females have deposited their eggs in a nest, the eggs are incubated by the *male* (Grzimek et al. 1972:84).

Thus, unlike most birds, tinamous are solitary and polygynous, and the typical male/female roles in incubation are reversed. The tinamou is therefore a model not only of bad social behavior, but in their breeding habits they exemplify what would be extremely undesirable reproduction habits if practiced by humans—inconstancy in mating and the abandonment of the children by the mother.

Before leaving the tinamous of the earth to describe those of the sky, we should mention one other unusual feature, the eggs of the tinamou. "[The eggs] may be green, turquoise-blue, purple, wine-red, slate-gray or a chocolate color, and they often have a purple or violet lustre" (Grzimek et al. 1972:85).

Tinamou eggs are like segments of a rainbow, oval in shape and placed in a nest. In the mythology of the Desana Indians, tinamous are believed to have been the sole survivors of a world fire and were responsible for preserving, through their eggs, all the colors of the rainbow (Reichel-Dolmatoff 1978b:280). One might say, then, that tinamous lay rainbows, and the dark cloud constellation of the Tinamou is located at the center of the arc of the Milky Way, which, as we have seen in the cosmology of Misminay, is considered equivalent to a nocturnal rainbow. Therefore, the example of the Tinamou elaborates an idea which appeared with the celestial Serpent (Mach'ácuay); that is, the equation of a celestial dark spot with the rainbow, an equivalence of black and multicolored.

As mentioned earlier, the dark cloud constellation of the Yutu (see fig. 65 and *SC*# 47) is located at the foot of the group of stars which, in Western astronomy, is known as the Southern Cross (Crux). Therefore, the astronomical periodicities of the Southern Cross will be virtually identical to those of the Yutu. The heliacal rise and set dates (September 3 and April 22, respectively) of the principal star of the Southern Cross, α Crucis, give a very close approximation of the agricultural season in the Andes. In addition, α Crucis and the Yutu transit the upper meridian on the morning of the December solstice sunrise, and on the morning of the June solstice they are transiting the lower meridian. Phrased another way, the Yutu is at its zenith point on the morning of the

December solstice, and it is at its nadir point on the morning of the June solstice. The periodicities of the Yutu therefore relate the celestial bird both to agriculture and to the solstices.

The mating season of tinamous in the Cuzco area is not reported in the literature. However, the breeding season of several varieties of tinamous in the northern hemisphere extends from February through April (Pearson 1955 and Friedmann 1950). This period, in the northern hemisphere, is related to the onset of longer days, warmer weather, and the approach of the rainy season following the December solstice. Alexander Skutch (1976:72) has shown that "birds transported across the equator seem to adapt their nesting to the season appropriate to their new environment." Thus, when translating the factors related to the breeding season of northern-hemisphere tinamous to the high Andes of the southern hemisphere (that is, the lengthening of the day and the onset of the warm/rainy season), we arrive at a period of a couple of months after the *June* solstice, July through early September. Another factor related to the beginning of breeding in birds is the increased availability of food, a condition met for the seed-eating tinamous not only by the beginning of the rainy season, but also by the August-September planting of seed crops such as maize.

Felipe Guamán Poma de Ayala, in his description of the various Incaic agricultural duties of the year, records the times when the crops must be guarded from birds and animals; the guard duties begin with the planting in late August and end with the harvest in early May. If we combine, in a single calendar, the times when Guamán Poma (1936:1130 ff.) mentions explicitly the need to guard the crops with the times of the heliacal rise and set dates of the celestial Tinamou, we arrive at the close correlation shown in figure 67.

By comparing this figure to the calendar of toads and agriculture (fig. 66) and to the data discussed earlier concerning the relationship of rainbow/serpents and the rainy season, we can understand the larger set of associations between agriculture, rainbows, and the celestial Tinamou. Rain, and therefore rainbows, occur in the southern Andes primarily during the period from September through April. Figure 67 shows that this is also the period when the celestial Tinamou is in the sky. The beginning of the calendrical correlation between the celestial Tinamou and the crops is also related to the breeding period of terrestrial tinamous, and the total span of the calendar (from September through late April) is the period when terrestrial tinamous represent a threat to agriculture.

Llama (*SC*# 24, 40, 43, 45)

Perhaps the most conspicuous dark cloud constellation—since it virtually fills the sky overhead during the rainy season—is the Llama (see fig. 65). We are in a considerably stronger position when we discuss the symbolic and ritual associa-

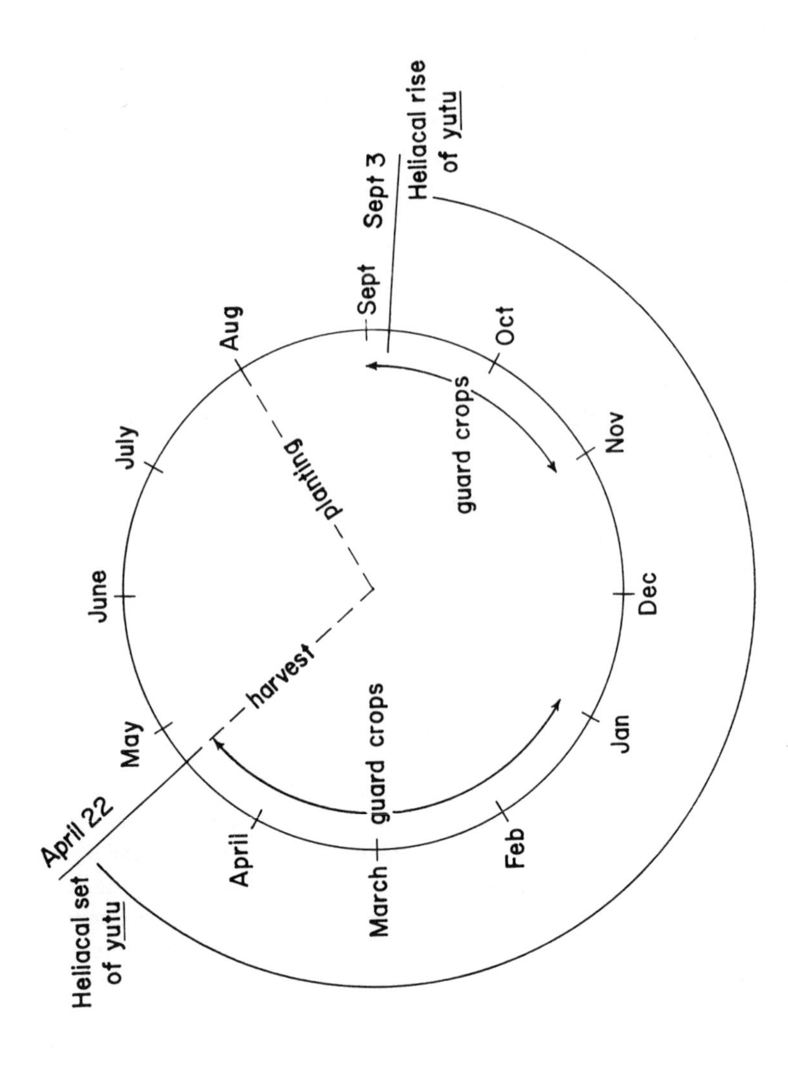

Fig. 67. A Calendar of Tinamous and Agricultural Guard Duties

tions of the dark cloud Llama because the Spanish chroniclers were much more explicit in their descriptions of this animal and its relation to Inca rituals (see Zuidema and Urton 1976 for a more complete discussion of the ethnohistorical material). Here the concern is largely with the relationship between constellations and the "procreation and augmentation" of animals on the earth, and discussion is confined to a comparison of the biological and astronomical cycles of terrestrial and celestial llamas.

As John Murra (1965) has demonstrated so well through the ethnohistorical documentation, the llama has for some time been an essential animal in Andean life. Always used as a beast of burden, it also provides meat for food, wool for clothing, and dung for fertilizer and is considered an appropriate gift to the gods in the form of a sacrifice.

The breeding period of llamas starts in late December, and the gestation period lasts eleven months. Llamas begin to give birth from late November to early December, with the birthing period ending in March. Llamas, and the closely related camellid alpaca, give birth between 6:00 A.M. and noon (Jorge Velasco N., personal communication, 1977). Thus, for Andean pastoralists, the early morning hours of the rainy season in December are important times for caring for newly born calves and for attempting to ward off predators such as the fox.

For the herder who rises early in the morning during the llama birthing period, an especially fortuitous sight will be the appearance, over the southeastern horizon, of the two bright stars, α and β Centaurii, which are referred to as *llamacñawin* ("the eyes of the llama," fig. 32 and *SC#* 24); they are the first part of the huge body of the llama to appear over the horizon. After the heliacal rising of the eyes of the Llama in late November, the eyes and the body rise progressively higher in the sky each morning until, in late April, at the end of the birthing season of llamas, the dark cloud constellation of the Llama stands along the north-south meridian at midnight.

Llamas were also incorporated in the Inca calendar system in the form of sacrifices made at fixed intervals during the agricultural season. Brown and brownish-red llamas were sacrificed from August to September at the beginning of the planting season. White llamas were sacrificed and black ones tied to a post and starved to death to induce rain and growth of the crops, and multicolored llamas were sacrificed at the time of the harvest in late April or early May (Polo de Ondegardo 1916:chap. 6).

The black llamas starved in October and the multicolored ones sacrificed in April-May are of special interest because the dates which they mark in the calendar define an important solar axis in the Inca calendar. It has been shown (Zuidema and Urton 1976:86) that the inferior culmination of α and β Centaurii at midnight on October 30 was coupled with the zenith sun (which occurs on the same day) as a way of fixing the solar dates in the Inca calendar for the initiation of young Inca nobles one month later in late November. The superior culmination of α and β Centaurii at midnight in April occurred on the same night that the sun stood in the nadir at midnight. Thus, the alternation of the superior and inferior culmination at midnight of the eyes of the Llama was seen in relation to the alternation of the zenith and nadir sun in October and April, and these dates were marked in the ritual cycle by the sacrifice of black and multicolored llamas. Francisco de Avila, who wrote an account of the beliefs and customs of the community of Huarochirí in the early seventeenth century, has given us a description of the dark cloud constellation of the Llama which explicitly refers to the descent of the celestial Llama (Yacana) to the earth at midnight. The Llama drinks the waters of the swollen rivers at the beginning of the rainy season, she is then shorn of her wool (which turns out to be multicolored despite the fact that Avila describes her as having the color of a shadow), and finally she returns to her position in the celestial River (Avila 1966:chap. 29). Therefore, in the case of the llama, we find the same theme of dark and multicolored symbolism which has appeared earlier, and we also find a further correlation of a dark cloud constellation with rain and the sun.

Atoq ("fox", *SC#* 35)

The South American fox (*Dusicyon culpaeus*) inhabits wooded, hilly country ranging up to four thousand meters in the Andes. Most of the six South American fox species are nocturnal hunters, which may contribute to their projection in the dark clouds of the night sky. The diet of the fox is omnivorous and includes birds, rabbits, frogs, toads, and an occasional sheep; in fact, the fox feeds on most of the fauna it pursues through the sky in Quechua astronomy. In addition, William L. Franklin, in a study of the social behavior of the vicuña (1974:486), reports that foxes prey both on adult and baby vicuña. The vicuña defense against attacks by foxes is "group mobbing." An excellent description of this appears in the seventeenth century chronicle of Bernabé Cobo:

> There are usually a large number of foxes where vicuñas live; and the foxes chase and eat the young of the vicuñas. The vicuñas defend their children in the following way. Many vicuñas rush together to attack the fox, striking it until it falls to the ground. They then run over it many times without giving it a chance to get up until they kill it by their blows. The cries of the miserable fox are useless as he succumbs to the feet of the vicuñas (Cobo 1964, vol. 1:368; my translation).

The position of the celestial Fox in relation to the baby Llama and the hind legs of the mother Llama (fig. 65) appears to fix in the clouds of interstellar dust in the Milky Way this well-recorded motif of pursuit and trampling. Like the tinamou, the fox also has a tendency to freeze when endangered. There is one account of such a "frozen" animal being approached by a man and remaining motionless even when struck with a whip handle (Walker et al. 1964, vol. 2: 1160).

The mating season of foxes falls in midwinter; in South America the season extends from late June through September (Ewer 1973:309). With a gestation period of around ten weeks, baby foxes generally appear from October through December. The relationship of foxes with the solstices was pointed out in chapter 3, where it was noted that in Misminay it is commonly believed that foxes give birth principally on one day of the year: December 25, four days after the solstice. In addition, the runa of Misminay pinpoint exactly the spot where baby foxes are born every year. It occurs on the side of Apu Wañumarka at a point which is precisely the setting point of the June solstice sun as viewed from the community. However, the solstitial relation of foxes goes beyond their birth near the time of the December solstice sunrise at the place of the June solstice sunset.

The dark cloud constellation of the Atoq (Fox) is a rather amorphous dark spot which stretches at a right angle from the tail of Scorpio crossing the ecliptic between the Western constellations of Scorpio and Sagittarius. The importance of this celestial position is that, as the sun travels along the path of the ecliptic throughout the year, it "enters" the constellation of the Fox at the time of the December solstice. Therefore, as the sun rises in the southeast with the constellation of the Fox around the time of the December solstice, terrestrial foxes are born on the earth in the antisolstitial direction (that is, in the direction of the June solstice sunset; see fig. 68).

To extend this solstitial/fox analysis further, the other passage of the sun through the Milky Way occurs at the time of the June solstice, the time when the sun sets on the side of Apu Wañumarka. Therefore, since the breeding season of foxes begins in late June and baby foxes are born in December, the life cycle of the fox is directly associated not only with the sun in its two solstice positions, but also with the times and places of the intersection of the sun with the celestial River, the Milky Way.

The Dark Clouds in Andean Cosmology

Aside from the close correlation between biological and astronomical phenomena in these data, the dark cloud animal constellations serve as the focus for a number of important classificatory and symbolic principles in the astronomy and cosmology of Misminay. Principal among these are color oppositions and associations (dark, light, and multicolored); concepts of fertility (such as the cosmic circulation of water, earthly fecundity [pachatira], animal procreation);

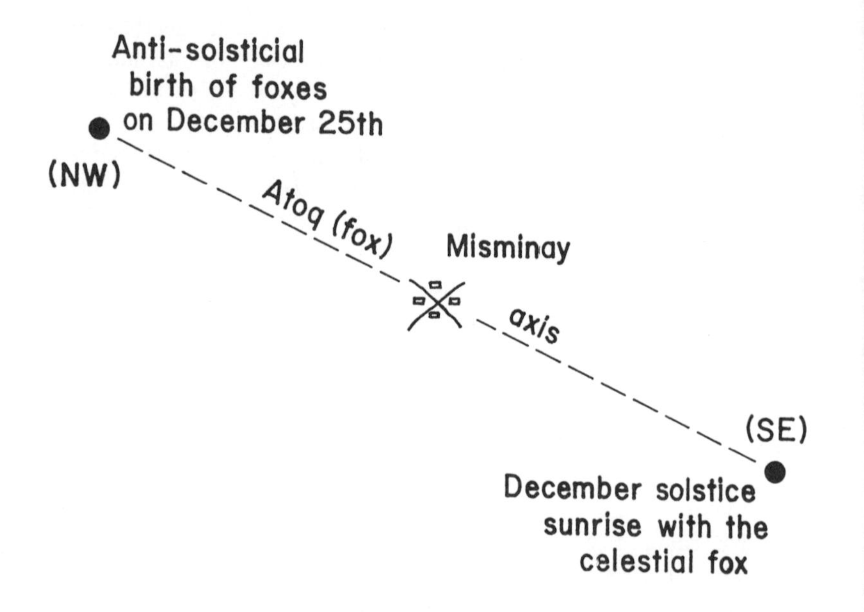

Fig. 68. The Axis of the Fox

and several fundamental principles of spatial and temporal orientation. The latter are seen primarily in the relationship between rainbows, the Milky Way (as a nocturnal celestial river), and the sun. The section on the Atoq (Fox) detailed a coincidence in the orientations of the sunrise points of the solstices, as viewed from the community of Misminay, with the orientations of the two axes of the Milky Way during the early evenings at the times of the solstices. These coincidences have been diagramed in figure 22. As illustrated there, on the evenings of the solstices the Milky Way is seen as an arc opposite the setting sun. This is exactly the same relationship that exists between rainbows and the sun; rainbows are always seen as arcs (or circles) stretching across the sky directly opposite the sun.

These observations suggest, then, a synchronized pattern of celestial and meteorological lines and points of orientation which are in constant motion but which retain, throughout the annual cycle, a persistent, internal pattern of oppositions. As the sun moves south, the arcs of rainbows and the Milky Way rotate slowly northward so as to maintain their opposition to the sun. This implies that the seasons are determined not by the sun alone but also by celestial arcs of a certain orientation at a certain time of the day, night, or year. In studies of tropical forest astronomical symbolism, Claude Lévi-Strauss (1978:110-113) has described a similar pattern of relationships among the sun, rainbows, the Milky Way, and the moon.

As a final point, it must be remarked that my study of the dark cloud constellations has brought us only to the threshold of another important problem which could be discussed in relation to the animals of Quechua astronomy—that is their metaphoric and symbolic importance in the cosmology of the Quechuas. This chapter takes a rather literal approach to the equation of specific dark clouds with specific animals. Essentially the equations are studied with regard to the information they provide concerning the correlation of units of space and time in the calendar system; I have not extensively investigated the *meanings* of the equations. In so doing, we have perhaps gained certain insights at the expense of others. The study of animal symbolism in South America and elsewhere has begun to occupy more of the attention of ethnologists, and it is a topic of fundamental importance in understanding how various cultures organize and process the information provided within their own natural environment. We need to understand, for instance, the particular cultural factors which determine the boundaries between wild and domesticated animals, as well as the symbolic associations of animals that depend on whether they are classified as aquatic, amphibious, terrestrial, or celestial; carnivorous, herbivorous, or omnivorous; feathered, furred, or scaled; and so forth. All these are features which will give us important clues concerning the metaphoric and symbolic place of animals in Incaic and contemporary Quechua culture. The present study of the animals of the astronomy of Misminay provides an initial perspective in the study of animals and culture in the Quechua universe, but it must be broadened to incorporate the role of animals in other domains of Quechua thought. At the least, it is hoped that the exegesis of one short sentence from the chronicle of Polo de Ondegardo has contributed toward an understanding of his observation that "in general, they believed that all the animals and birds on the earth had their likeness in the sky in whose responsibility was their procreation and augmentation" (Polo 1916:chap. 1).

10. Summary and Conclusions

In the Introduction, I stated that my principal aims in undertaking ethnoastronomical fieldwork in Misminay were to investigate the extent to which contemporary astronomical beliefs and practices in the Andes operate as a coherent system and to determine the relevance of the contemporary astronomical data in the study of Incaic astronomy, cosmology, and calendrics. How do the data from Misminay serve in realizing these objectives?

The Nature of Contemporary Andean Astronomy

When we discuss the systematic or integrative nature of the astronomical knowledge in contemporary Quechua communities, we are responding to those few sources in the literature which have, until now, guided our thinking on this question. In 1940, after presenting a description of astronomical beliefs in the community of Kauri, Bernard Mishkin (1940:241) concluded that the astronomical and cosmological ideas "are today fragmentary, unsystematic and are not brought into relation to ritual or to the everyday life of Kauri." In fact, the only consistent context within which astronomical data have been discussed has been the relation between celestial bodies, the agricultural cycle, and crop predictions (see Casaverde Rojas 1970, Lira 1946, Mishkin 1940, and Tuero Villa 1973).

In the years since Mishkin's study of cosmological beliefs in the southern Andes, our comparative knowledge of contemporary Quechua cosmology and our thinking about it have undergone a virtual revolution (e.g., Bastien 1973, Earls 1972, Gow 1976a, Isbell 1978, Mayer 1974, Urbano 1974, Barrette 1972, Morisette and Racine 1973, Vallée 1972, and Wagner 1976 and 1978). However, the relative sophistication with which we are now able to discuss Quechua "terrestrial" cosmology has not been matched by an increase in our knowledge and understanding of the astronomy. We have developed a disembodied view of the Quechua universe; it is a view of cosmology which has its feet planted firmly on the ground and its head lost in a celestial chaos.

One consequence of our present view of Quechua cosmology is that discussions of Quechua notions of time and space center on successive periods (or ages) of *historical/mythical* time and on *terrestrial* space. In order to flesh out our understanding of cosmology in the Andes, we must reintegrate *celestial* space with terrestrial space and *operational* time (i.e., astronomical, biological, and seasonal periodicities) with historical time. The present study has demonstrated that, although some of the pre-Conquest astronomical knowledge may have been submerged or syncretized with Western European concepts, contemporary Quechua astronomical beliefs are not, as Mishkin and others would have us believe, "fragmentary and unsystematic."

In the first place, astronomical data are far richer than a general review of the ethnographic literature might indicate. In Misminay alone there are some forty different named stars and constellations and a large body of data relating the stars and constellations to the sun, the moon, and the planets. But more importantly, the astronomical data are not disconnected, nor are they lacking a cultural, economic, or mythological context.

In this regard, astronomy and cosmology are initially of interest because of their relation to the primary concerns of Quechua society, specifically, subsistence. In Misminay, subsistence is a concept inseparable from the fertility of crops and animals. Therefore, we may suppose that the varying degrees of interest which the runa of Misminay direct toward the natural world will be concerned with this ultimate concern for survival through an efficient, integrated system of agriculture and pastoralism. Emphasis in this study has been on those several areas in which the system of astronomy is directly related to the growth of the crops and the reproduction of animals (e.g., the center/planting sun; crops planted in accordance with the phases of the moon; the relation between the Pleiades, the tail of Scorpio, and the crops; and the correlation of biological and astronomical cycles found in the discussion of the dark cloud animal constellations). However, in a number of other areas, such as the chapters dealing with the celestial crosses and the stars of the twilight, no specific relationship to subsistence activities is indicated. This is perhaps owing to a lack of data, a problem which can be remedied through additional fieldwork; however, it should be pointed out that these "nonsubsistence" chapters have provided a more than rudimentary understanding of the structure and organization of local calendar systems in the Cuzco area. One requirement of an efficient agricultural system is the ability to predict climatic changes and to organize social and individual activities with respect to these changing conditions; therefore, our increased understanding of the calendar system gives us a better understanding of the units and divisions of time which provide the framework for the sequence of agricultural tasks, the pastoral cycle, and other ritual and social activity cycles.

In Misminay, we also found that celestial phenomena are integrated into the cosmology in a consistent and coherent manner. This is seen in the organization of the celestial sphere by opposing points along intercardinal axes, in the consistent opposition between astronomical and cosmological orientations along one intercardinal axis (southeast-northwest) with those along the other intercardinal axis (southwest-northeast), in the use of the Milky Way as a circle or plane of orientation for the divisions of celestial space, and in the systematic correlation of biological and astronomical periodicities.

In Misminay, the divisions of time are directly related to the divisions of space; the most obvious example of this is the cycling of the seasons. The rainy season and the dry season are related to variations in the orientation not only of the sun, but of the Milky Way as well. The periods or phases of the moon, which are used to time the planting of the crops, have been shown to have a relation to the general principle of classifying units of celestial space with respect to the position of the sun. Therefore, the structural and organizational properties which are projected onto the sky result in a systematic integration of celestial bodies and cycles with the calendar of activities.

The orientation of the axes for the division of terrestrial space in Misminay is shown to closely correspond to the orientation of the axis of the division of celestial space. In effect, the two principal axes of the Milky Way in the zenith coincide with the two principal footpath/irrigation-canal axes in Misminay. An immediate objection to this conclusion might be that the terrestrial axes are determined by topographic features and especially by the orientation of the flow of water. However, the importance of this coincidence has been argued not just on the basis of the specific orientations of the footpaths but, more importantly, also on the basis of a number of consistent relationships that are made between named topographic (e.g., horizon) features and named celestial features. To point out the consistency between the two sets of axes which articulate these features in their respective domains is not necessarily to suggest a cause-and-effect relationship; rather, to do so merely notes the consistency of one more aspect of the total organization.

However, in view of the fact that this particular feature of Quechua cosmology has been developed on the basis of the topography of the community of Misminay, future fieldwork must address two questions. First, if orientations within the community of Misminay are found to be determined by the relationship between certain terrestrial and celestial features, then how will the different topography of another community be reflected in that community's organization and structuring of celestial space? That is, will the quadripartitioning of community space always mirror intercardinal axes established by celestial features such as the zenith axes of the Milky Way—or will there be a selection of different celestial features, ones which mirror the principal topographic orientations within that particular community? Second, are the celestial phenomena observed in crop predictions and animal reproduction standard from community to community? If so, how are they observed in relation to different horizon lines (i.e., different topographies)?

Observations on the astronomical system in the community of Misminay call for a complete reassessment of our present notions of astronomy and cosmology (i.e., of the nature and status of precise knowledge) in the Peruvian Andes. The most urgent changes which must be effected are related to our assumptions concerning the complexity and sophistication of the Quechua formalization of astronomical knowledge and the degree to which the celestial bodies, periodicities, and orientations are integrated with the organization of terrestrial space.

Incaic and Contemporary Quechua Astronomy

Throughout this study, we have seen that the integration of celestial phennomena with cosmological structures and everyday activities is a complex subject requiring a complete understanding of local topography, agricultural cycles, biological periodicities, and several other aspects of the specific celestial and terrestrial environment within a given community. Any attempt to briefly summarize some of the principal similarities between Incaic and contemporary Andean astronomy must be done on a somewhat superficial level, for a complete integration could only be accomplished by correlating the topographic organization within a single Incaic community (e.g., the ceque system of Cuzco) with the ethnoastronomical data recorded within that same community. Obviously, this task would require at least another volume. Therefore, the following summary is intended only to bring together some of the more general lines of evidence which suggest a strong similarity between Incaic and contemporary Quechua astronomical knowledge and practices.

The sun is still referred to by the name used in Inca times (*inti*), and the belief that each Inca was the "son of the sun" is reflected in contemporary references to the sun as *Manco Capac* and *Huayna Capac* (the names of the first and eleventh Incas). The Incas divided the movement of the sun along the horizon into units which represented monthly periods in the calendar (see Cobo 1964: 158 and Zuidema's discussion, 1977a:220), and the sun was also observed to determine the time of planting (Anonymous 1906:151). In Misminay, the "three sections of the sun" may represent a division of the area traversed by the sun which is derived from or based on a calendrical principle similar to the Incaic practice of establishing schematic solar/lunar time periods. I have suggested that the solar observations made by the Incas to determine the time of planting in the Cuzco Valley are similar in many respects to the solar horizon calendar now used in Misminay for timing agricultural periods and duties. Furthermore, in both cases such observations may be related to the spatial and temporal boundaries defined by the zenith and antizenith (= nadir) sunrise points. Whereas the observations at the beginning of the planting season in Inca Cuzco were made in August by viewing the sunrise/sunset in relation to pillars (sucancas) placed along the horizon, they are now made in Misminay by observing the sunrise from a named place along the horizon (i.e., its rise from the mountains above Calca), and by the celebration of the Catholic saints' days during August and September, which closely coincide with the Calca sunrise.

We know from the chronicles that the solstices were two of the most important annual celebrations in the Inca ritual calendar. In Misminay, the solstice observances have been syncretized with the Catholic celebrations of Navidad (December 25) and San Juan (June 24). Beyond this, in Misminay the four points of the solstices along the horizon (SE, SW, NW, NE) are related to the points of intersection of the extensions of the two intercardinal terrestrial axes with the centers of the quarters of the Milky Way. Therefore, we must reexamine the ethnoastronomical data provided by the Spanish chroniclers relating to the Milky Way, the solstices, and the zenith and nadir sun in order to determine whether, in Inca cosmology, the alternating intercardinal axes of the Milky Way were integrated with important solar rise and set points. This will then need to be tested against, for example, the formal organization of the topography of the valley of Cuzco via the ceque system. These data and suggestions tell us that the sun has been, and continues to be, one of the principal celestial bodies for ordering units of time and space in the Quechua universe.

As in the case of solar terminology, the moon (quilla) is called by the same name today as during Inca times, and it is referred to by a title which relates it to the Inca female nobility (Colla and/or Coya Capac). Synodic and perhaps sidereal lunar periods were used as the basis of the Inca ritual calendar and for the regulation of agricultural duties (see Anonymous 1906:140-152, Guamán Poma 1936:fol. 235 and 260, and Zuidema 1977a:238-250). In Misminay, since many of the annual agricultural duties extend over periods of about a month, these may well be related to an indigenous lunar ritual, agricultural, and pastoral calendar. In addition, the lunar phases are used in different ways with respect to the planting of different crops. The specific phase used depends on whether the crop produces its edible parts above or below ground. Because we have seen a correlation between the Incaic terminology for the lunar phases and the terminology used today for the morning, evening, and midnight zenith stars (e.g., *ch'issin* and *pacarin*), we may suppose that Incaic astronomy and calendrics were also characterized by the integration of lunar and stellar timing mechanisms with the observation of the sun by means of horizon pillars.

The most important aspect of lunar astronomy and symbolism, its relation to women and the menstrual cycle, remains largely unstudied. This is the one area of ethnoastronomical research which has the most potential for significantly increasing our understanding of the Quechua and Inca calendar system and of determining the possible existence of a standardized female lunar "zodiac."

In the final pages of this study, I must make an admission of bias, and it is, indeed, a "stellar" bias. At this point, of course, my admission comes as no surprise, for at least one third of this work has been concerned with analyzing stellar data. Before the contemporary and Incaic stellar material is compared, I feel it is important to explain the origin of this bias and its impact on my research.

To begin with, my stellar bias is the result of a "prebias," or a prior assumption on my part, which can be stated as follows: when Francisco Pizarro arrived on the shores of Tahuantinsuyu in 1532, the Incas possessed an astronomical and cosmological system which was as complex and sophisticated as any that existed at that time anywhere in the world. The men who destroyed the Inca empire seem, by comparison, superstitious, rude barbarians who stood at the vanguard of an expanding, militaristic, and technologically superior continent which had achieved very little significant advance in its own system of astronomy and cosmology since the time of Aristotle and Ptolemy. It must be remembered that the *De Revolutionibus* of Copernicus, which became known to the public (but not published) by about 1531, had little significant impact on Western cosmological thinking until almost a half century after Copernicus's death in 1543. As one historian of Western science has noted, "In the Western world . . . there was little to record for nearly five centuries after Ptolemy. After that time ensued an almost total blank, and several more centuries elapsed before there was any appreciable revival of the interest once felt in astronomy" (Berry 1961:76).

To take one particularly relevant example, Stanley L. Jaki (1975) has amply documented the fact that Aristotle's pronouncement on the Milky Way, that it was composed of a substance akin to "swamp gas" raised to a level in the sky directly below the sphere of the fixed stars,[1] was effectively the most persuasive view of the galaxy in Europe until the early seventeenth century. Aristotle's theory of the Milky Way was, of course, primarily concerned with the problem of the substance out of which the Milky Way was formed, but the critical point is that from the time of Aristotle until the day Columbus set sail very little was said in European cosmology about the complex motion of the ring of the Milky Way through the sky as a possible source of order in the universe (see Berry 1961:33 and Jaki 1975:93). I contend that we have found in the community of Misminay an understanding and a systematization of celestial (and especially galactic) motions which is *at least* as profound as anything that existed in Europe until the invention of the telescope in the first decade of the seventeenth century.

Having stated the above heretical prebias, I can now proceed to the functional bias: my concern with the stars. The only way to effectively demonstrate the relative sophistication of both Incaic and contemporary Quechua astronomy and cosmology is to determine whether or not they are in "control" of the night sky. For instance, did the extremely high level of bureaucratic, sociocultural, and spatial organization employed in the administration of the Inca empire stop at the horizon? That such may have been the case is, in fact, an absurd proposition, and it is for this reason that I earlier criticized John H. Rowe's view of the nature of Incaic calendrical correlations as totally unsystematic and Bernard Mishkin's statements regarding astronomy and cosmology in Kauri.

There is ample evidence, both in the ethnohistorical documentation and in contemporary ethnographic studies, that the Quechua-speaking Indians of the Andes have long had a cosmological interest in the sun. In fact, if anything, we have been dazzled by the "sun kingdoms of the Americas" and consequently have blinded ourselves to the true depth and scope of the total system. Only in recent years, primarily through the studies of R. T. Zuidema (especially 1977a), have we been made aware of the importance of the moon in the ritualism, symbolism, and calendar system of the Incas. However, until now we have had only the slightest evidence that the stars are incorporated in a meaningful and systematic way into a general cosmological structure. Unless we can demonstrate that the stellar and planetary cycles are integrated with the solar and lunar cycles, we cannot refute with authority the persistent suggestion that the pre-Columbian Incas and the contemporary Quechua-speakers live(d) in anything but the kind of primordial, mythical chaos which certain of the chroniclers would have us believe reigned in Peru prior to the time of the Inca empire (e.g., Anonymous 1906:149-151). Because I do not believe that people anywhere will live for very long in chaos, I suggest that we assume that complex systems of astronomy and cosmology have existed in the Peruvian Andes since at least Chavín times (i.e., since about 1200 B.C.; see Lathrap's discussion of Chavín iconography [1977] for evidence of the existence of a well-integrated Chavinoid agricultural and cosmological system). Given these several assumptions and biases, what has the present study contributed to our understanding of this central issue—and what remains to be investigated?

In the first place, one area in which much more intensive fieldwork is needed concerns the planets and the status of planetary cycles and symbolism in the system of cosmology. Chapter 8 put forward a number of specific suggestions for the spatial and temporal framework within which a study of the planets may proceed. Aside from studies dealing with the relationship of women and the moon, the investigation of the planets is one of our most pressing ethnoastronomical concerns.

As to a judgment concerning the contribution of this work to an understanding of the similarity between contemporary Quechua and Incaic stellar astronomy, I will begin by naming the Incaic stars, constellations, and dark nebulae which can now be identified in the sky. Of some thirty-five Incaic stars and/or constellations compiled from various ethnohistorical sources, the following sixteen are identified (with varying degrees of confidence) in the star catalogue (table 7):

#4	Ch'issin (ch'aska)	González 1952
#7	Illarimi (ch'aska)	González 1952
#9	Pachapacariq ch'aska	("pacarikchaska") González 1952 and Guamán Poma 1936:fol. 895, 885
#16	Collca (Oncoy)	Avila 1966:chap. 29, Arriaga 1920, Cobo 1964:159 and Polo 1916:3
#17 and 36	Contor	Avila 1966:chap. 29
#18	Chakána	Cobo 1964:160 and Polo 1916:5
#19	Choquechinchay	González 1952 and Pachacuti Yamqui 1950:226
#24	Llamacñawin	Avila 1966:chap. 29
#26	Mirco Mamana (Mamana micuc)	Cobo 1964:160 and Polo 1916:5
#30 and 31	Pisqa Collca (Coyllur)	Avila 1966:chap. 29
#40	Llama (and Urquchillay)	Avila 1966:chap. 29, Guamán Poma 1936:fol. 895, 885, and Polo 1916: 3-4
#41	Mach'ácuay	Cobo 1964:159 and Polo 1916:5
#42	Mayu	Cobo 1964:160
#43	Ombligo de la llama	Cobo 1964:159 and Polo 1916:4
#45	Uñallamacha	Cobo 1964:159 and Polo 1916:4
#47	Yutu	Avila 1966:chap. 29 and Guamán Poma 1936:fol. 895, 885

The list contains single bright stars, star-to-star constellations, and dark cloud constellations—virtually the total inventory of celestial forms and groupings recognized in Misminay and in several other present-day communities. In addition, an entire chapter is devoted to stellar crop predictions and divinations, and it was pointed out there that the contemporary crop predictions made by observing the Pleiades are identical to the predictions described by Francisco de Avila in the community of Huarochirí at the end of the sixteenth century (Avila 1966: chap. 29). We have also seen that Juan Polo de Ondegardo's statement attesting to the Incaic belief in a close correlation between celestial and terrestrial animals has been borne out by examination of the identifications and periodocities of the dark cloud animal constellations (chap. 9).

However, in the analysis of stellar material, focus has been on the system of celestial orientations based on the Milky Way. We now know how this system operates in the community of Misminay; the following discusses evidence of its existence in Inca times in order to suggest how the concept of the Milky Way arrived at here may be of use in the study of Incaic cosmology and symbolism.

I have elsewhere suggested that one of the lines of the four suyu (quarter) divisions of the ceque system of Cuzco appears to have been related to an intercardinal, Milky Way orientation (Urton 1978b). The dividing line between the quarters of Cuntisuyu and Collasuyu extended from the center of Cuzco (i.e., from the Temple of the Sun) toward the southeast, where it intersected the line of the horizon at the point of the rising of α Crucis and the Coalsack (Yutu). This point is important because it is within the area along the Milky Way which falls nearest to the unmarked south celestial pole and therefore revolves around it. Thus, in the Quechua Milky-Way-based system of celestial orientations, this point, or general area, represents the southern limit of the movement of the celestial waters which flow through the sky within the Mayu (the celestial River).

This suggests, in turn, that the southeast-northwest line was an important orientational and cosmological axis in Inca Cuzco. In fact, we have ethnohistorical confirmation of the ritual importance of the SE-NW axis in the chronicle of Cristóbal de Molina (1916). Molina describes an annual ceremonial pilgrimage made during the June solstice by the Inca priests. The route traveled by the priests took them straight from Cuzco toward the southeast to the temple of Vilcanota, where, in Inca mythology, the sun was believed to have been born. This site of the birth of the sun is known today by the name of La Raya (see Duviols 1978). At La Raya, the priests reversed their direction and returned to Cuzco by walking northwestward, but now along the Vilcanota River (Molina 1916:26-27 and Zuidema 1977c and 1978b:349-350). I suggest that the route traveled by the priests from Cuzco to the southeast was thought of as more than a terrestrial pilgrimage; it was equivalent to a walk along the Milky Way to the point of terminus and origin of the universe. The return to Cuzco from the southeast, from the place of origin of the sun (the sun = the Inca), was an annual ritual of regeneration of the Inca and a reincorporation of the sun into the ritual, calendrical, and cosmological organization of the empire. But the journey to La Raya along the Vilcanota River involves more than the sun and the Inca for it also involves a reenactment of the creation of the universe by Viracocha, the creator god of the Incas.

In several places throughout this study it is stated that the Vilcanota River, which flows from the southeast to the northwest (as viewed from Cuzco), is equated with the Milky Way. In the Incaic myth of the creation of the world (see the collection of creation myths in Pease 1973), the creator god, Viracocha ("sea fat" or "foam of the sea"), rises out of Lake Titicaca, passes through the sky, and goes across the sea at Manta, in Ecuador.

Study of the route traveled by Viracocha shows that from the point of view of Cuzco, Viracocha travels from the southeast to the northwest; this is also, we know, the orientation both of the Vilcanota River and of one of the two principal axes of the Milky Way. Thus, it is reasonable to suppose that Viracocha, the creator god of the Incas, was equated with the Milky Way and the Vilcanota River—that is, with the two cosmic rivers which are connected at the edge of the earth. From Misminay, we have explicit testimony concerning the means whereby the terrestrial waters are taken up by the Milky Way and circulated through it. We have an equally explicit but less complete explanation of this same process from the seventeenth-century chronicle of Bernabé Cobo (1964:160): "They say that through the center of the sky there crosses a great river which is marked by that white belt which we see from here below called the Milky Way . . . they believed that this river took up the water which flows across the earth" (my translation).

If we now look again at the cosmological drawing of Pachacuti Yamqui (fig. 69), we see in the upper central portion of the drawing a large ellipse which Pachacuti uses as the symbol of Viracocha. This symbolic image of the Incaic creator god probably represents the ring, or more properly the ellipse, of the Milky Way and its terrestrial reflection, the Vilcanota River.[2]

This interpretation of the symbol used by Pachacuti is supported by ethnographic data from the community of Tomanga in the central Andes. At the time of the cleaning of the irrigation canals in February, the people of Tomanga construct a large altar in the plaza (fig. 70). The arch stretching over the altar is referred to by the term *Pusuqu* ("foam"). E. Pinto Ramos (1970:174) describes the Pusuqu as an "arch of boughs covered with a white construction composed of linked ovals." In addition, he says that *pusuqu* is the name given to small oval objects placed on the four altars dedicated to the four irrigation canals.

The term *pusuqu* has appeared elsewhere in the present study. An informant in Misminay described the Milky Way as composed of two rivers which originate in the north, arch through the sky in opposite directions, and collide in the south near α Crucis; the "foam" of their collision is called *pusuqu*. Thus, the foam which results from the union of celestial rivers is referred to in Misminay by the same term that is used in Tomanga for the arch and the oval as symbols of the unification of terrestrial rivers; therefore, the *union* and the *foam* can be symbolized by the arch, the oval, or, as in Pachacuti's drawing, the ellipse. Note at this point that foam, and moving water in general, are equated in Andean symbolism with semen, the masculine force of fertilization (see Isbell 1978:143 and

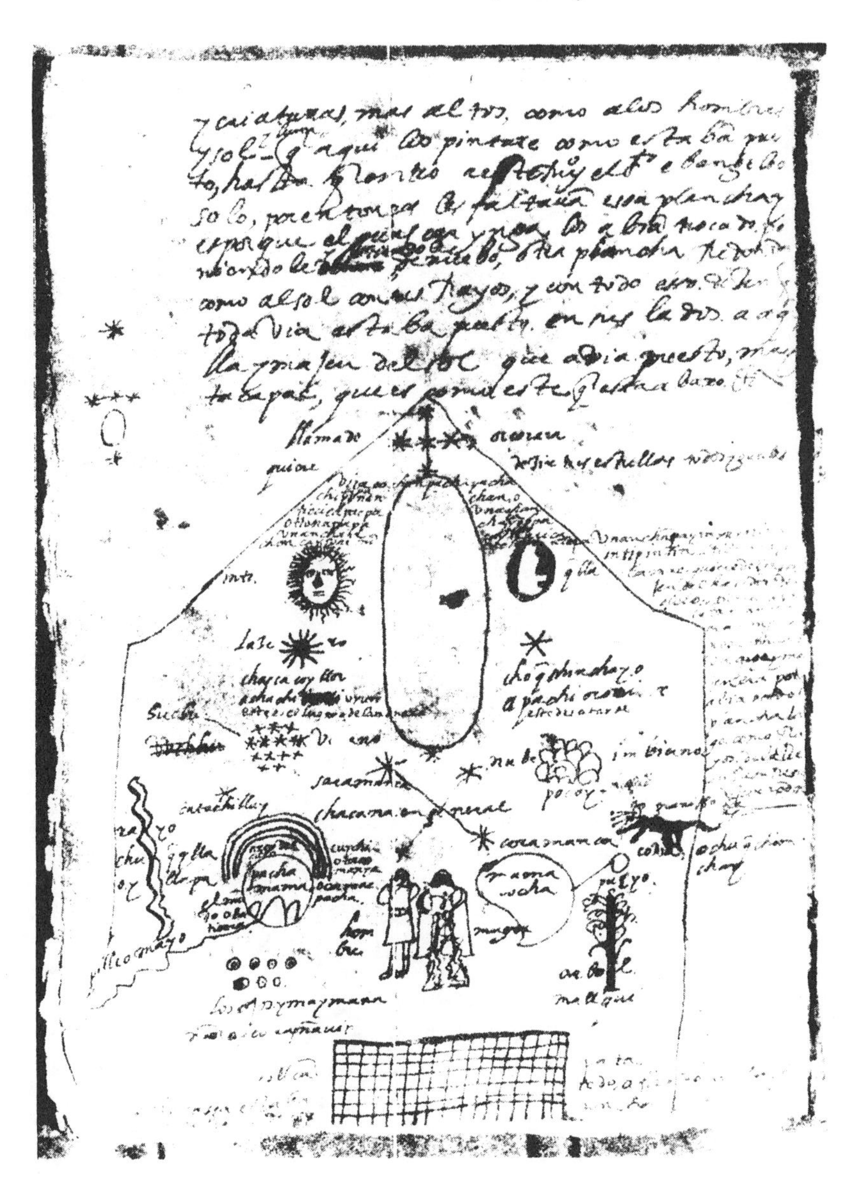

Fig. 69. The Cosmological Drawing of Pachacuti Yamqui

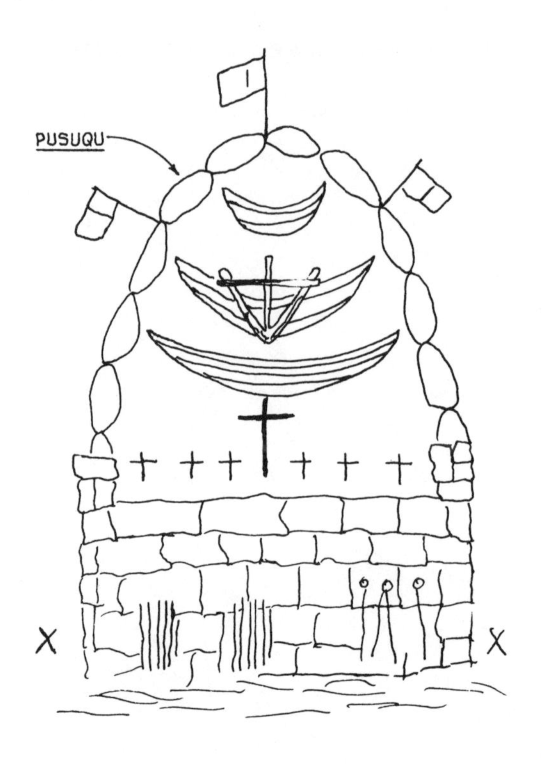

Fig. 70. The Irrigation-Canal Cleaning Altar in Tomanga

164 and Ossio 1978:381). Viracocha ("sea fat" or "sea foam"), the creator god of the Incas, is the synthesis of opposing motions or objects both in the sky (the Milky Way) and on the earth (the Vilcanota River).

In Bernabé Cobo's description of Incaic cosmology written in 1653, he outlined the importance of the Milky Way in the short sentence quoted earlier (p. 202). Between the beginning and the end of that sentence, however, Cobo inserts the judgment that, concerning the Milky Way, the Incas "imagine a world of nonsense which would take too long to tell about" (Cobo 1964:160). It is my thesis that what we have found concerning the Milky Way in the community of Misminay is similar to what Cobo heard over three hundred years ago—and dismissed as a "world of nonsense."

Appendix

APPENDIX

STELLAR IDENTIFICATIONS REPORTED BY OTHER CUZCO-AREA ETHNOGRAPHERS

Bonett 1970:71	
a. *arado* ("plow")–located in the Milky Way. Used in the July crop predictions	*SC*# 12
Casaverde Rojas 1970:168	
a. *pacha paqary ch'aska* ("dawn of the earth/time star")–Venus or morning star	*SC*# 9
b. *ch'aska* ("shaggy hair"; star)–Orion or evening star	*SC*# 4
c. *Hatun Cruz* ("great cross")–Southern Cross	*SC*# 20, 21
d. *Qolqa* ("storehouse")–Ursa Major	*SC*# 16, 32
e. *Lluthu (Yutu* = "tinamou")–nebulosity	*SC*# 47
f. *Llamaq Ñawin* ("eyes of the Llama")–No I.D.	*SC*# 24
Lira 1946:18-19	
a. *Kkoto* ("pile")–No I.D. Used in the crop predictions of August	*SC* *SC*# 16, 32
b. *Cruz* ("cross")–Southern Cross	*SC*# 21
c. *Las Tres Marías* ("the three Marys")–the Belt of Orion; used in August predictions	*SC*# 18
d. *Sacos de Carbôn* ("sacks of carbon")–located within the Milky Way; used in August crop predictions	*SC*# 47
Mishkin 1940:240	
a. Pleiades (no name)–used in crop predictions	*SC*# 16, 32
b. Orion (no name)–". . . the form of a plow and oxen." Used in crop predictions	*SC*# 12, 18
B. S. Orlove, personal communication, 1978	
a. *Las Tres Marías* ("the three Marys")–Belt of Orion	*SC*# 18

Note: Names and identifications are those given by the ethnographers; *SC* numbers refer to related identifications in the star catalogue of table 7.

b. *Huchuy Cruz* ("small cross")–No I.D. *SC#* 21
c. *Wikuña*–Scorpio *SC#* 11, 25, 40
d. *Yutu* ("tinamou")–Coalsack *SC#* 47
e. *Mayu* ("river")–Milky Way; the denser sections of the Mayu = high population areas along a major river *SC#* 42
f. Shooting star–each one represents a cattle thief (*abigeo*) stealing an animal *SC#* 2, 3

Percy Paz F., personal communication to R. T. Zuidema, 1976
a. *Alqo* ("dog")–same position as *atoq* ("fox") *SC#* 35
b. *Vicuña*–beneath Llama *SC#* 40, 45
c. *Señoracha* ("Mother Mary")–behind *alqo/atoq* *SC#* 49

Christopher Wallis, personal communication, 1976
a. Orion's Belt–head of *Wakaya* ("cow"?) *SC#* 18
b. The three stars of Orion's Belt = *Toro, Wajran, Ismalia* *SC#* 18
c. *Llama*–East of the Southern Cross *SC#* 40
d. *Paqocha* ("little priest")–8 to 10 stars perhaps near Orion's Belt
e. *Oveja* ("sheep")–No I.D.
f. *Alqo* ("dog")–No I.D. *SC#* 35?
g. *Mishi* ("cat")–No I.D.
h. *Puma*–No I.D.
i. *Sapo* ("toad")–a *yana phuyu* ("dark cloud") in the Milky Way *SC#* 37
j. *Yuthu* ("tinamou")–behind the *sapo* *SC#* 47
k. *Mayu* ("river")–Milky Way *SC#* 42
l. *Llama Ñawin* ("eyes of the llama")–No I.D. *SC#* 24
m. *Contor*–located in the southern skies *SC#* 17, 36
n. *Wikuña*–No I.D. *SC#* 40?
o. *Waman* ("falcon")–No I.D.
p. Man and dog driving a *Wikuña*
q. *Qhoto* ("pile")–six stars (Pleiades?) *SC#* 16, 32

William Sullivan 1979:67
a. *Wajus qana* (?); *qasa* ("frost")–Large Magellanic Cloud *SC#* 50?
b. *alto qana* ("high"); *holq'e* ("tadpole")–Small Magellanic Cloud *SC#* 49-50?
c. *cruz* ("cross"); *huch'uy cruz* ("small cross"); *huch'uy cruz calvario* ("small cross of calvary"); *lluthu cruz* ("tinamou cross")–Southern Cross *SC#* 21, 34
d. *lluthu* ("tinamou"); *cuntur* ("condor")–coalsack *SC#* 47, 36
e. *hanp'atu* ("toad")–black spot near Southern Cross *SC#* 37
f. *Llamaq ñamin* ("eyes of the llama")–α and β Centaurii *SC#* 24

g. *Mayu* ("river")	
ñan ("road")	(see chap. 2)
h. *Llama* ("llama")–black part of Milky Way from ϵ Scorpii to α and β Centaurii	*SC*# 40
i. *qoto* ("pile"); *qolqa* ("storehouse")–Pleiades	*SC*# 16, 32
j. *Cruz Calvario*–λ, ν, Zeta 1 and 2, mu 1 and 2, and ϵ Scorpii	*SC*# 15, 25, 28
Tuero Villa 1973:76-77	
a. *arado* ("plow")–located in the Milky Way. Used in the July crop predictions	*SC*# 12
b. *Llama*–located in the Milky Way	*SC*# 40

Notes

Introduction

1. Elizabeth Baity (1973:390) defines *ethnoastronomy* as the merging of "astronomy, textual scholarship, ethnology, and the interpretation of ancient iconography for the purpose of reconstructing lifeways, astronomical techniques, and rituals." This definition will serve to describe in general what I mean by the term *ethnoastronomy*, although I have used the terms *astronomy* and *ethnoastronomy* interchangeably in this study.

2. See Georges Posener (1965) for a discussion of the concepts of up/down and right/left in relation to north and south in Egyptian cosmological orientations. D. F. Pocock (1965) has studied the symbolism of north and south in early Judeo-Christian thought.

3. Archaeoastronomy is "the analysis of the orientations and measurements of megalithic and other monumental ancient structures, many of which . . . could have served for the prediction of solar and lunar eclipses and unquestionably *did* serve for the determination of solstices and equinoxes, enabling the setting of dates for agricultural activities and for the ritual cycle of the year (Baity 1973:390)."

4. Deborah A. Poole (n.d.) has argued convincingly that the ceque system was also used as the basis for the wider, territorial organization of the Inca empire and that we may find remnants of this organizational structure in contemporary Andean pilgrimage routes and circuits.

5. Cf. Gary Urton 1978a and 1979. As this manuscript was being readied for publication, I received a copy of William F. Sullivan's thesis submitted for the M. Litt. degree at the University of St. Andrews (1979). Sullivan's thesis, "Quechua Star Names," is based on five months of ethnoastronomical fieldwork in northern Bolivia and southern Peru. I have not extensively incorporated Sullivan's data in the text of this book, but a list of the astronomical phenomena he identified appears in the Appendix. The list shows that Sullivan's data and identifications accord perfectly with those which I found in the course of my

own work in southern Peru. Sullivan's thesis is an important addition to our comparative knowledge and understanding of Quechua astronomy, and it is hoped that his manuscript, or portions therefrom, will soon appear in published form.

6. Patricio Arroyo Medina's discussion of the various groups inhabiting the plain of Maras is rather confusing. He first states that he is quoting local tradition ("de la boca de mis antepasados") and then, before listing the names of the groups, he quotes Carlos E. Bárcena Cruz, *Historia del Perú*. Unfortunately, I do not have access to Bárcena's work.

Chapter 2: The Organization and Structure of Space

1. From 1975 to 1977 ethnoastronomical data were collected in the following communities, all of which are in the Department of Cuzco (see map 1):

a. *Yucay* (District of Urubamba). Population 2,000. Yucay is located three kilometers upriver from Urubamba.

b. *Sonqo* (District of Qolqepata). A small dispersed community some seventy-nine kilometers northeast of Cuzco (see Wagner 1978).

c. *Quispihuara* (District of Santa Ana). Population ca. 25. Located a few kilometers northwest of Cuzco.

d. *Lucre* (District of Oropesa). Population 2,000. Lucre is situated at the mouth of a small river valley (across the valley from the pre-Incaic archaeological site of Pikillaqta) about thirty kilometers southeast of Cuzco.

2. Informants are quite clear about the distinction between the vulture (*contor*) and the beings in otra nación (*condores*).

3. Catherine Wagner (1978:84-99) has described a similar organization of houses into named clusters in the community of Sonqo. Each of the named clusters in Sonqo is said to be a separate ayllu (localized social group).

4. It is instructive to note the similarities between the Quechua conception of the Milky Way and that found in traditional Chinese cosmology: "The celestial river divides into two branches near the north pole and runs from there to the south pole. One of the branches passes near the asterism Nan-teou (Lambda Sagittarii), and the other near the asterism Toungtsing (Gemini). The river is the celestial water which flows through the sky and falls on the earth" (Schlegel 1967:20; my translation).

5. In ancient China also, the Milky Way was believed to be connected to the rivers of the earth. The Milky Way is the "means whereby the deities communicate with the Four Quarters" (Schafer 1977:259-260). This is also the case in India, where the Milky Way is considered to be a heavenly river, the Celestial Ganges (Kulkarni 1962:37). In the *Vishnu Purana*, the Ganges is said to be born of the Milky Way (Santillana and von Dechend 1969:259). For a comparative discussion of the celestial river in Greek and Sanskrit literature, see Scharfe 1972.

Chapter 3: The Sun and the Moon

1. An informant of David D. Gow's from the area of Ocongate stated that

the first alpacas were female and black (Gow 1976a:195). This may be significant in relation to not only the figure seen in the lunar spots, but also the presence of a large dark cloud constellation which is referred to as a female llama (*SC*# 40).

Chapter 4: Meteorological Lore

1. For a good description and analysis of Incaic meteorological/genealological symbolism, see Mariscotti de Görlitz (1973).

2. Juvenal Casaverde Rojas (1970:171) also gives two "classes" of rainbows, but he does not mention a classification according to sex. The two types are *k'uychi* ("rainbow") and *wankar k'uychi* ("small drum rainbow").

3. One modern meteorological study has shown that haloes around the sun or moon are accurate portents of rain at least seven times out of ten (Thompson 1961).

Chapter 5: The Stars and Constellations

1. The substitution for Venus of another star or planet as the morning or evening star or both was also a characteristic of ancient Assyrian astronomy. Hildegard Lewy (1965:274) writes, "As Venus, even as all the other moving stars, has well-determined periods of invisibility, the ancients substituted for it a fixed star of similar appearance to which the worshipper could turn during Venus' absence."

Chapter 6: Collca: The Celestial Storehouse

1. The heliacal rise of a star is taken to be the first day on which the star is visible in the east before dawn; the heliacal set is considered as being the first day on which a star sets in the west before dawn. All azimuth readings, as well as heliacal rise and set dates, are taken from computer read-out charts provided by Dr. Anthony Aveni, Colgate University.

2. The phenomenon of the alternating replacement and terminological equation of opposing stars has also been noted in Coptic and Indian astronomy (Santillana and von Dechend 1969:361).

3. It should be pointed out here that I was not present when all of the observations and predictions were carried out. In most cases, the information represents informants' descriptions of how Collca *should* be observed for predictive purposes.

4. The description was obtained from a young woman (about twenty years old) from Quispihuara. Women are responsible for a number of agricultural duties, but agriculture is not the area of their primary interest nor responsibility.

5. If the new rather than the full moon is considered, the relationship is reversed: on the morning of June 24 the tail of Scorpio will set at sunrise with a full moon, but it will *rise* at *sunset* with a new moon.

6. Juan Tuero Villa (1973:94) identifies Collca as Ursa Major, the Big Dipper. This identification is interesting because Ursa Major can only be seen for some four months (from mid-April through mid-August) at the latitude of

Cuzco; thus, it will be visible, low on the northern horizon, during the time of the Collca observations described in this chapter. However, since there is no supporting evidence for this identification (not even from Jorge Bonett Yépez), I will not at this time attempt to analyze the role of Ursa Major as a possible referent of Collca.

Chapter 7: Crosses in the Astronomy and Cosmology of Misminay

1. Material from this chapter has appeared in article form as "Celestial Crosses: The Cruciform in Quechua Astronomy" in the *Journal of Latin American Lore* 6, no. 1 (1980).

2. The word *linun* is from the Latin *lignum* ("wood"). The term refers to the actual wood of the cross but it was used very early in Christian literature as a translation of the Greek ξυλον, which refers to crucifixion on a wooden cross (Reijners 1965:6-8). Thus, we can translate Linun Cruz either as the "wooden cross" or as the "cross of crucifixion."

Chapter 8: The Stars of the Twilight

1. In Cailloma, Christopher Wallis found that the after-sunset radiance is called Antayllupi ("copper hair"). From November to March, Antayllupi is said to cause abortions in animals; this effect can only be counteracted by *kachi rumi* ("salt rock"; C. Wallis, personal communication, 1976).

2. In his study entitled *Primitive Time Reckoning,* Martin Nilsson (1920: 170) found only one instance of the use of lunar *quarters.* Most lunar calendar systems are based on three phases: new moon, full moon, and old crescent.

3. One informant in Misminay identified the "star of the north" as Altair, the "star of the west" as Canopus, and the "star of the south" as Sirius. The identifications, however, were made on a star map. I have decided to center the discussion of the four pachapacariq ch'askas on the orientational data in figure 60 rather than on identifications from a star map.

4. The heliacal rise and set dates in table 14 are for latitude -13° (the latitude of Misminay is ca. $-13^\circ 30'$) at A.D. 1500.

Chapter 9: Yana Phuyu: The Dark Cloud Animal Constellations

1. A version of this chapter has appeared in article form as "Animals and Astronomy in the Quechua Universe" in *Proceedings of the American Philosophical Society* 125, 2 (April 1981).

Chapter 10: Summary and Conclusions

1. The following is a summary description of Aristotle's theory of the Milky Way.

"[Aristotle] waxed unusually verbose in expounding the details of his own explanation of the Milky Way. If it had any logical implication, it was that the Milky Way should be constantly changing both in shape and in shade. First, there had to be a steady efflux of the dry exhalation from the marshy regions of the earth. Then, this had to rise at a steady rate across the turbulences of the atmosphere many thousands of miles into a specific, ring-shaped volume of space with most intricate contours, located directly under that belt of the sphere of fixed stars where stars were most numerous. Once there, the given volume of dry exhalation had to maintain the same density and spread to secure the same rate of incandescence. This slow burning was in turn caused, according to Aristotle, by frictional heat from the daily revolution of the great number of stars in the belt of the Milky Way" (Jaki 1975:5).

2. From this identification of the Milky Way with the symbol for Viracocha drawn by Pachacuti Yamqui, we arrive at an Incaic conception of the shape of the Milky Way which is very like that postulated by John Kepler: "Kepler asked his reader to consider the earth as though it were far out of the plane determined by the Milky Way. In that case the Milky Way should appear in one single look as a very small circle or an ellipse visible in its entirety, whereas now only one half of it can be seen at any moment" (Jaki 1975:108).

Bibliography

Acosta, José de

1954 *Historia natural y moral de las indias* [1590]. Biblioteca de Autores Españoles, vol. 73. Madrid: Ediciones Atlas.

Aguilar Páez, Rafael

1970 *Gramática quechua y vocabularios.* Lima: Universidad Nacional Mayor de San Marcos.

Albó, Javier

1972 "Dinámica en la estructura inter-comunitaria de Jesús de Machaca." *América Indígena* 32, no. 3:773-816.

Anonymous

1906 "Discurso de la sucesión y gobierno de los Yngas." *Juicio de límites entre el Perú y Bolivia*, edited by Víctor Maurtua, 8:149-165.

Aranguren Paz, Angélica

1975 "Las creencias y ritos mágico-religiosos de los pastores puneños." *Allpanchis Phuturinqa* 8:103-132.

Arguedas, José María

1956 "Puquio, una cultura en proceso de cambio." *Revista del Museo Nacional* (Lima) 25:184-232.

Arriaga, Pablo José de

1920 *La extirpación de la idolatría en el Perú* [1621]. Colección de Libros y Documentos Referentes a la Historia del Perú, edited by Horacio H. Urteaga. 2nd series, vol. 1. Lima.

Arroyo Medina, Patricio

1974 "Estudio del complejo arqueológico de Moray como fuente de enseñanza-aprendizaje de la historia y experimentación agrícola." Thesis presented to the Faculty of Education, Universidad Nacional de San Antonio Abad del Cuzco.

Ascher, Marcia, and Robert Ascher

1972 "Numbers and Relations from Ancient Andean Quipus." *Archive for the History of Exact Sciences* 8:288-320.

1975 "The Quipu as a Visible Language." *Visible Language* 9:329-356.

1978 *Code of the Quipu Databook*. Ann Arbor: University of Michigan Press. (Ann Arbor, Mich.: University Microfilms International, #78-68652.)

n.d. "Code of the Quipu: A Study in Media, Mathematics and Culture." Ms.

Aveni, Anthony F.

1972 "Astronomical Tables Intended for Use in Astro-Archaeological Studies." *American Antiquity* 37, no. 4:531-540.

n.d. a "Horizon Astronomy in Incaic Cuzco." In *Archaeoastronomy in the Americas*, edited by Ray Williamson. Santa Barbara: Ballena Press. In press.

n.d. b "Reply to Rowe's Review: 'Archaeoastronomy in Mesoamerica and Peru.'" *Latin American Research Review*, in press.

Aveni, Anthony F., Horst Hartung, and Beth Buckingham

1978 "The Pecked Cross Symbol in Ancient Mesoamerica." *Science* 202, no. 4365:267-279.

Avila, Francisco de

1966 *Dioses y hombres de Huarochirí* [1608]. Translated by J. M. Arguedas). Lima: Instituto de Estudios Peruanos.

Baity, Elizabeth Chesley

1973 "Archaeoastronomy and Ethnoastronomy So Far." *Current Anthropology* 14:389-449.

Barrette, Christian

1972 "Aspects de l'ethno-écologie d'un village andin." *Canadian Review of Sociology and Anthropology* 9, no. 3:255-267.

Barthel, T. S.

1971 "Viracochas Prunkgewand." *Tribus* 20.

Bastien, Joseph W.

1973 *Qollahuaya Rituals: An Ethnographic Account of the Symbolic Relations of Man and Land in an Andean Village.* Cornell University Latin American Studies Program Dissertation Series, no. 56. Ithaca, N.Y.

Becher, Hans

1976 "Moon and Reincarnation: Anthropogenesis as Imagined by the Surára and Pakidái Indians of Northwestern Brazil." In *The Realm of the Extrahuman: Ideas and Actions*, edited by Agehananda Bharati, pp. 337-345. The Hague: Mouton.

Berry, Arthur

1961 *A Short History of Astronomy from Earliest Times through the Nineteenth Century.* New York: Dover.

Betanzos, Juan diez de

1924 *Suma y narración de los Incas* [1551]. Colección de Libros y Documentos Referentes a la Historia del Perú, edited by Horacio H. Ur-

teaga, 8:79-208. Lima: Imprenta y Librería San Martí y Ca.
Blair, W. Frank, ed.
1972 *Evolution in the Genus Bufo.* Austin: University of Texas Press.
Blake, Emmet R.
1977 *Manual of Neotropical Birds.* Vol. 1. Chicago: University of Chicago Press.
Bolton, Ralph
1976 "Andean Coca Chewing: A Metabolic Perspective." *American Anthropologist* 78:630-633.
Bonett Yépez, Jorge W.
1970 "La agricultura en una comunidad indígena del Cuzco." Thesis, Universidad Nacional San Antonio Abad del Cuzco.
Boulenger, George A.
1961 *Catalogue of the Snakes in the British Museum* (original publication, 1896, in 3 vols.), vol. 2. New York: Hafner Pub. Co.
Boyer, Carl B.
1959 *The Rainbow from Myth to Mathematics.* New York: T. Yoseloff.
Brownrigg, Leslie Ann
1973 "A Model of the Andean System of Time: Dispersed and Congregated Ritual Calendars." Paper read at the seventy-second meeting of the American Anthropological Association, New Orleans (symposium on Andean Time: Ritual Calendars and Agricultural Cycles).
Brush, Stephen
1975 "Parentesco y agricultura en un pueblo andino peruano." *América Indígena* 35, no. 2:367-389.
1977 "Kinship and Land Use in a Northern Sierra Community." In *Andean Kinship and Marriage*, edited by Ralph Bolton and Enrique Mayer. American Anthropological Association, Special Publication, no. 7.
Cadogan, León
1973 "Some Plants and Animals in Guaraní and Guayakí Mythology." In *Paraguay: Ecological Essays*, edited by J. Richard Gorham, pp. 97-104. Miami: Academy of the Arts and Sciences of the Americas.
Callegari, G. V.
1914 "Conoscenze astronomiche degli antichi Peruviani." *Rivista Abruzzese* 29, no. 3:113-126.
Carrión Cachot, Rebeca
1955 *El culto al agua en el antiguo Perú.* Lima: Tipografía Peruana (separata de la *Revista del Museo Nacional de Antropología y Arqueología* 2, no. 2.)
Casaverde Rojas, Juvenal
1970 "El mundo sobrenatural en una comunidad." *Allpanchis Phuturinqa* 2:121-243.

Cayón Armelia, Edgardo
1971 "El hombre y los animales en la cultura quechua." *Allpanchis Phuturinqa* 3:135-162.
Cei, José M.
1972 "*Bufo* of South America." In *Evolution in the Genus Bufo,* edited by W. Frank Blair. Austin: University of Texas Press.
Cobo, Bernabé
1964 *Historia del nuevo mundo* [1653]. Biblioteca de Autores Españoles, vol. 92. Madrid: Ediciones Atlas.
Dalle, Luis
1969 "El despacho." *Allpanchis Phuturinqa* 1:139-154.
1971 "Mosoq Wata, Año Nuevo." *Allpanchis Phuturinqa* 3:34-44.
de Schauensee, Rodolphe Meyer
1970 *A Guide to the Birds of South America.* Wynnewood, Penn.: Livingston.
Ditmars, Raymond L.
1937 *Snakes of the World.* New York: Macmillan.
Doughty, Paul L.
1967 "La cultura, la bebida y el trabajo en un distrito mestizo andino." *América Indígena* 27, no. 4:667-687.
DuGourcq, Jean
1893 "L'Astronomie chez les Incas." *Revue Scientifique* 52:265-272.
Duviols, Pierre
1966 "La Visite des idolatries de Concepcion de Chupas (Pérou, 1614)." *Journal de la Société des Américanistes* (Paris) 55:497-510.
1967 "Une Inedite par Cristóbal de Albornoz: 'La instrucción para descobrir todas las huacas del Piru y sus camayos y haciendas'." *Journal de la Société des Américanistes* (Paris) 56:7-39.
1976 "Une Petite chronique retrouvees." In *Errores, ritos, supersticiones y ceremonias de los yndios de la provincia de Chinchaycocha y otras del Piru,* edition and commentary by P. Duviols. *Journal de la Société des Américanistes* (Paris) vol. 63.
1978 "Un Symbolisme andin du double: La lithomorphose de l'ancêtre." *Actes du XLIIe Congrès International des Américanistes* (Paris) 4: 359-364.
Earls, John
1971 "The Structure of Modern Andean Social Categories." *Journal of the Steward Anthropological Society* 3, no. 1.
1972 "Andean Continuum Cosmology." Ph.D. dissertation, University of Illinois, Urbana-Champaign.
1973a "La organización del poder en la mitología quechua." In *Ideología mesiánica del mundo andino*, edited by Juan Ossio. Lima: Instituto de Estudios Peruanos.
1973b "Long Term Social Periodicities and the Ceque System as a Computer." Paper read at the seventy-second meeting of the American

Anthropological Association, New Orleans (symposium on Andean Time: Ritual Calendars and Agricultural Cycles).

1976 "Evolución de la administración ecológica inca." *Revista del Museo Nacional* (Lima) 42:207-245.

1979 "Astronomía y ecología la sincronización alimenticia del maíz." *Allpanchis Phuturinqa* 13:117-135.

Earls, John, and Irene Silverblatt

1978 "La realidad física y social en la cosmología andina." *Actes du XLIIe Congrès International des Américanistes* (Paris) 4:299-325.

Escobar, Gabriel

1973 *Sicaya: cambios culturales en una comunidad mestiza andina.* Lima: Instituto de Estudios Peruanos.

Ewer, R. F.

1973 *The Carnivores.* Ithaca, N.Y.: Cornell University Press.

Ferguson, George Wells

1904 *Signs and Symbols in Christian Art.* New York: Oxford University Press.

Fioravanti, Antoinette

1973 "Reciprocidad y economía de mercado." *Allpanchis Phuturinqa* 5:121-130.

Fitch, Henry S.

1970 *Reproductive Cycles in Lizards and Snakes.* University of Kansas, Museum of Natural History, Miscellaneous Publications No. 52.

Fock, Neils

n.d. "Ecology and Mind in an Andean Irrigation Culture." Ms.

Fonseca Martel, César

1976 "Organización dual del sistema en las comunidades de Chaupiwaranga, Perú." *Actas del XLI Congreso Internacional de Americanistas* (Mexico City) 3:545-552.

Foster, George M.

1960 *Culture and Conquest: America's Spanish Heritage.* Viking Fund Publications in Anthropology, edited by S. L. Washburn, no. 27.

Franklin, William L.

1974 "The Social Behavior of the Vicuña." In *The Behavior of Ungulates and Its Relation to Management*, vol. 1, edited by V. Geist and F. Walther. Morges, Switzerland: International Union for Conservation of Nature and Natural Resources.

Friedmann, Herbert, and Foster D. Smith, Jr.

1950 "A Contribution to the Ornithology of Northeast Venezuela." *Smithsonian Institution, Proceedings of the United States National Museum,* vol. 100, no. 3268. Washington, D.C.

Gade, Daniel W.

1972 "Comercio y colonización en la zona de contacto entre la sierra y las tierras bajas del valle del Urubamba en el Perú. *XXXIX Congreso Internacional de Americanistas* (Lima) 4:207-221 (Actas y Memorias).

Gaposchkin, Sergei

1960 "The Visual Milky Way." In *Vistas in Astronomy*, edited by Arthur Beer, 3:289-295.

Garcilaso de la Vega, El Inca

1966 *Royal Commentaries of the Incas* (1609), part 1. Translated by Harold V. Livermore. Austin: University of Texas Press.

González Holguín, Diego

1952 *Vocabulario de la lengua general de todo el Perú llamada lengua QQuichua o del Inca* [1608]. Lima: Instituto de Historia, Universidad Nacional Mayor de San Marcos.

Gow, David D.

1974 "Taytacha Qoyllur Rit'i." *Allpanchis Phuturinqa* 7:49-100.

1976a "The Gods and Social Change in the High Andes." Ph.D. dissertation, University of Wisconsin, Madison.

1976b "Verticality and Andean Cosmology: Quadripartition, Opposition, and Mediation." *Actes du XLIIe Congrès International des Américanistes* (Paris) 4:199-211.

Gow, David D., and Rosalinda Gow

1975 "La alpaca en el mito y el ritual." *Allpanchis Phuturinqa* 8:141-174.

Gow, Rosalind, and Bernabé Condori

1976 *Kay Pacha*. Cuzco. Centro de Estudios Rurales Andinos "Bartolomé de las Casas."

Graham, F. Lanier (ed.)

1975 *The Rainbow Book*. Berkeley: The Fine Arts Museums of San Francisco.

Grzimek, Bernard, ed.

1972 *Grzimek's Animal Life Encyclopedia*. Vol. 7, *Birds I*. New York: Van Nostrand Reinhold Co.

1974 *Grzimek's Animal Life Encyclopedia*. Vol. 5, *Fishes II, Amphibians*. New York: Van Nostrand Reinhold Co.

Guamán Poma de Ayala, Felipe

1936 *El primer nueva crónica y buen gobierno* [1584-1614]. Travaux et memoires de l'Institut d'Ethnologie, 23. Paris: Université de Paris.

Guardia Mayorga, César A.

1971 *Diccionario Kechwa-Castellano, Castellano-Kechwa*. Lima: Editora Los Andes.

Hagar, Stansbury

1902 "The Peruvian Star-Chart of Salcamayhua." *Congrès International des Américanistes* 12:272-284.

1905 "Cuzco, the Celestial City." *Acts of the International Congress of Americanists* 13:217-225.

Hawkins, Gerald

1968 "Astro-archaeology." In *Vistas in Astronomy*, edited by Arthur Beer, 10:45-88.

Isbell, Billie Jean

1973 "Andean Structures and Activities: Towards a Study of Transformations of Traditional Concepts in a Central Highland Peruvian Community." Ph.D. dissertation, University of Illinois, Urbana-Champaign.

1974 "Parentesco andino y reciprocidad. Kuyaq: los que nos aman." In *Reciprocidad e intercambio en los Andes peruanos*, edited by Giorgio Alberti and Enrique Mayer, pp. 110-152. Lima: Instituto de Estudios Peruanos.

1976 "La otra mitad esencial: Un estudio de complementariedad sexual andina." In "La Mujer en los Andes." 5, *Estudios Andinos año* 5, vol. 5 no. 1:37-56.

1977 "Those Who Love Me." An Analysis of Andean Kinship and Reciprocity within a Ritual Context." In *Andean Kinship and Marriage,* edited by Ralph Bolton and Enrique Mayer, pp. 81-105. American Anthropological Association, special publications, no. 7.

1978 *To Defend Ourselves: Ecology and Ritual in an Andean Village.* Latin American Monographs, no. 47, Institute of Latin American Studies. Austin: University of Texas Press.

Jaki, Stanley L.

1975 *The Milky Way: An Elusive Road for Science.* New York: Science History Publications.

Jara, Victoria de la

1975 *Introducción al estudio de la escritura de los Inkas.* Lima: Instituto Nacional de Investigación y Desarrollo de la Educación.

Kelly, David H.

1960 "Calendar Animals and Deities." *Southwestern Journal of Anthropology* 3:317-337.

Knowlton, Frank H.

1909 *Birds of the World.* New York: H. Holt and Co.

Kulkarni, B. R.

1962 *Astronomical Origin of the Hindu Trinity.* Nander, India: Godateer Itihas Sanshodhan Mandal.

Lagercrantz, Sture

1952 "The Milky Way in Africa." *Ethnos*, vols. 1-4, pp. 64-72.

Lancaster, D. A.

1964 "Life History of the Boucard Tinamou in British Honduras," parts 1 and 2. *The Condor* 66, no. 3:165-181 and no. 4:253-276.

Larrea, Juan

1960 "*Lihuis* pajareros." *Corona Incaica.* University Nacional de Córdoba, Argentina.

Lathrap, Donald W.

1977 "Gifts of the Cayman: Some Thoughts on the Subsistence Basis of Chavin." In *Pre-Columbian Art History,* edited by Cordy-Collins and Stern, pp. 333-351. Palo Alto, Calif.: Peek Publications.

Leach, Edmund R.
1950 "Primitive Calendars." *Oceania* 20, no. 4:245-262.
Lehmann-Nitsche, Robert
1928 "Coricancha." *Revista del Museo de La Plata* 7, series 3:1-256.
Lévi-Strauss, Claude
1973 *From Honey to Ashes.* New York: Harper and Row.
1978 *The Origin of Table Manners.* London: Harper and Row.
Lewy, Hildegard
1965 "Ištar-Sâd and the Bow Star," In *Studies in Honor of Benno Landsberger on his Seventy-Fifth Birthday.* Assyriological Studies, no. 16. Chicago: The Oriental Institute of the University of Chicago.
Lipkind, William
1940 "Carajá Cosmography." *Journal of American Folklore* 53:248-251.
Lira, Jorge
1946 *Farmacopea tradicional indígena y prácticas rituales.* Lima: Talleres Gráficos "el Condor."
n.d. *Diccionario Kkechuwa Español.* Cuzco: Librería León.
Lumbreras, Luis G.
1969 "Acerca del desarrollo cultural en los andes." In *Mesa Redonda de Ciencias Prehistóricas y Antropológicas.* 2:125-154. Lima: Instituto Riva-Agüero.
Maegraith, B. G.
1932 "The Astronomy of the Aranda and Luritja Tribes." *Royal Society of South Australia, Transactions* 56:19-26.
Mariscotti de Görlitz, Ana María
1973 "La posición del señor de los fenómenos meteorológicos en los panteones regionales de los Andes Centrales." *Historia y Cultura* 6:207-215.
Mayer, Enrique
1974 "Más allá de la familia nuclear." *Revista del Museo Nacional* (Lima) 40:303-333.
Minnaert, M.
1954 *The Nature of Light and Colour in the Open Air.* New York: Dover.
Mishkin, Bernard
1940 "Cosmological Ideas among the Indians of the Southern Andes." *Journal of American Folklore* 53:225-241.
Mitchell, William P.
1977 "Irrigation Farming in the Andes: Evolutionary Implications." In *Peasant Livelihood*, edited by R. Halperin and J. Dow, pp. 36-59. New York: St. Martin's Press.
Molina, Cristóbal de (el Cuzqueño)
1916 *Relación de las fábulas y ritos de los Incas* [1573]. Colección de Libros y Documentos Referentes a la Historia del Perú, vol. 1, edited by Horacio H. Urteaga. Lima.

Morissette, Jacques, and Luc Racine
1973 "La Hierarchie des Wamani: Essai sur la pensée classificatorée Quechua." *Signes et Langages des Amériques* 3, nos. 1-2:167-188.

Morote Best, Efraín
1955 "La Fiesta de San Juan, el Bautista." *Archivos Peruanos de Folklore* 1; no. 1:160-200.

Mountford, Charles P.
1978 "The Rainbow Serpent Myths of Australia." In *The Rainbow Serpent,* edited by Ira A. Buchler and Kenneth Maddock, pp. 23-97. The Hague and Paris: Mouton.

Müller, Rolf
1929 "Die Intiwatana (Sonnenwarten) in alten Peru." *Baessler-Archiv* 13, nos. 3-4:178-187.
1972 *Sonne, Mond und Sterne über dem Reich der Inka.* Berlin: Springer Press.

Murra, John V.
1965 "Herds and Herders in the Inca State." In *Man, Culture and Animals,* edited by Anthony Leeds and Andrew P. Vayda. Publication no. 78 of the American Association for the Advancement of Science, Washington, D.C.
1972 "El control vertical' de un máximo de pisos ecológicos en la economía de las sociedades andinas." In *Visita de la provincia de León de Huánuco* [1562], *Iñigo Ortiz de Zúñiga Visitador,* pp. 429-476. Huánuco, Peru: Universidad Nacional Hermilio Valdizán.

Nachtigall, Horst
1975 "Ofrendas de llamas en la vida ceremonial de los pastores." *Allpanchis Phuturinqa* 8:133-140.

Nilsson, Martin P.
1920 *Primitive Time Reckoning.* Lund, Sweden.

Nimuendajú, Curt
1948 "The Mura and Piraha." *Handbook of South American Indians* 3: 255-269. Washington, D.C.: Bureau of American Ethnology.

Noble, G. Kingsley
1931 *The Biology of the Amphibia.* New York: McGraw-Hill.

Nordenskiöld, Erland
1925 "Calculations with Years and Months in the Peruvian Quipus." *Comparative Ethnographical Studies* (Göteborg, Sweden) 6, part 2: 1-35.

Núñez del Prado, Juan V.
1970 "El mundo sobrenatural de los Quechuas del Sur del Perú, a través de la comunidad de Qotobamba." *Allpanchis Phuturinqa* 2:57-119.

Núñez del Prado, Oscar
1973 *Kuyo Chico.* Chicago: University of Chicago Press.

Nussenzveig, H. Moyses
1977 "The Theory of the Rainbow." *Scientific American* 236:116-127.

Oesch, Will A.
1954 "Sobre algunos nombres populares del arco-iris." *Tradición, Revista Peruana de Cultura* 6:2-6.
O'Phelan Godoy, Scarlett
1977 "Cuzco 1777: El movimiento de Maras, Urubamba." *Histórica* 1, no. 1:113-128.
Orlove, Benjamin S.
1979 "Two Rituals and Three Hypotheses: An Examination of Solstice Divination in Southern Highland Peru." *Anthropological Quarterly* 52, no. 2:86-98.
Ortiz Rescaniere, Alejandro
1973 *De Adaneva a Inkarri.* Lima: Retablo de Papel Ediciones.
Ossio, Juan M.
1978 "El simbolismo del agua y la representación del tiempo y el espacio en la fiesta de la acequia de la comunidad de Andamarca." *Actes du XLIIe Congrès International des Américanistes* (Paris) 4:377-396.
Pachacuti Yamqui Salcamaygua, Juan de Santa Cruz
1950 *Relación de antigüedades deste reyno del Perú* [ca. 1613]. Reprinted in *Tres relaciones peruanas.* Asunción: Editorial Guarania.
Palomino Flores, Salvador
1968 "La cruz en los Andes." *Amaru* 8:63-66.
1971 "Duality in the Socio-cultural Organization of Several Andean Populations." *Folk* 13:65-88.
Pannekoek, A.
1929 "Een merkwaardig Javaansch sterrenbeeld." *Tijdschrift voor Indische Taal-, Land- en Volkenkunde: Bataviaasch Genootschap* 69: 51-55 (with a Note by Stein Callenfels, pp. 56-57).
Pearson, A. K.
1955 "Natural History and Breeding Behavior of the Tinamou, *Nothoprocta ornata.*" *The Auk* 72:113-127.
Pease, Franklin
1973 *El Dios Creador Andino.* Lima: Mosca Azul Editores.
Pinto Ramos, E.
1970 "Estructura y función en la comunidad de Tomanga." Thesis, Universidad Nacional de "San Cristóbal" de Huamanga.
Pocock, D. F.
1975 "North and South in the Book of Genesis." In *Studies in Social Anthropology: Essays in Memory of E. E. Evans-Pritchard,* edited by J. H. M. Beattie and R. G. Lienhardt, pp. 273-284. Oxford: Clarendon Press.
Polo de Ondegardo, Juan
1916 *Los errores y supersticiones de los indios sacados del tratado y averiguación que hizo el Licenciado Polo* [1571]. Colección de Libros y Documentos Referentes a la Historia del Perú, vol. 3. Lima: Im-

prenta y Librería San Martí y Ca.

Poole, Deborah A.
n.d. "Geography and Sacred Space in the Andean Pilgrimage Tradition." Ms.

Posener, Georges
1965 "Sur l'orientation et l'ordre des points cardinaux chez les Egyptiens." *Nachrichten der Akademie Wissenschaften in Göttingen Philologisch-Historiche Klasse*, no. 2:69-78.

Quiroga, Adán
1942 *La cruz en América.* Buenos Aires: Editorial Americana.

Radicati di Primeglio, Carlos
1965 "La 'seriación' como posible clave para descifrar los quipus extranumerales." *Documenta* 4:112-215.

Rauh, James H.
1971 "Tentative Reconstruction of the Peruvian Calendar System." Paper read at the thirty-sixth annual meeting of the Society for American Archaeology, Norman, Oklahoma.

Ravines Sánchez, Roger
1963-1964 "Ocho cuentos del zorro." *Folklore* 11-12, nos. 11-12:103-112.

Reichel-Dolmatoff, Gerardo
1975 *The Shaman and the Jaguar.* Philadelphia: Temple University Press.
1978a "The Loom of Life: A Kogi Principle of Integration." *Journal of Latin American Lore* 4, no. 1:5-27.
1978b "Desana Animal Categories, Food Restrictions, and the Concept of Color Energies." *Journal of Latin American Lore* 4, no. 2:243-291.

Reijners, Gerardus Q.
1965 *The Terminology of the Holy Cross in Early Christian Literature.* Nijmegen-Utrecht: Dekker and Van de Vegt N.V.

Roca W., Demetrio
1966 "El sapo, la culebra y la rana en el folklore actual de la pampa de Anta." *Folklore, Revista de Cultura Tradicional* 1, no. 1:41-66.

Rodríguez, Rivera, Virginia
1965 "El Arcoiris." *Folklore Americano* 13:52-69.

Roe, Nicholas A., and William E. Rees
1979 "Notes on the Puna Avifauna of Azangaro Province, Department of Puno, Southern Peru." *The Auk* 96:475-482.

Rowe, John H.
1946 "Inca Culture at the Time of the Spanish Conquest." *Handbook of South American Indians* 2:183-330. Washington, D.C.: Bureau of American Ethnology.
1979 "Archaeoastronomy in Mesoamerica and Peru." *Latin American Research Review* 14, no. 2:227-233.

Sallnow, Michael J.
1974 "La peregrinación andina." *Allpanchis Phuturinqa* 7:101-142.

Santillana, Giorgia de, and Hertha von Dechend
1969 *Hamlet's Mill: An Essay on Myth and the Frame of Time.* Boston: Gambit Incorporated.
Sarmiento de Gamboa, Pedro
1942 *Historia de los Incas* [1572]. Buenos Aires: Emecé Editores.
Schafer, Edward H.
1977 *Pacing the Void: T'ang Approaches to the Stars.* Berkeley: University of California Press.
Scharfe, Hartmut
1972 "The Sacred Water of the Ganges and the Styx-water." *Zeitschrift für Vergleichende Sprachforschung* 86:116-120.
Schaumberger-Gars, J.
1952 "Die Ziqpu-Gestirne nach neuen Keilschrifttexten." *Zeitschrift für Assyriologie* 50 (n.s. 16):214-229.
Schiller, Gertrud
1972 *Iconography of Christian Art.* Vol. 2, translated by Janet Seligman. Greenwich, Conn.: New York Graphic Society in Greenwich Connecticut.
Schlegel, Gustaf
1967 *L'Uranographic chinoise.* Taipei: Ch'eng-Wen Pub. Co.
Schmidt, Karl P., and Warren F. Walker, Jr.
1943a "Snakes of the Peruvian Coastal Region." *Zoological Series of the Field Museum of Natural History* 24, no. 27:297-324.
1943b "Three New Snakes from the Peruvian Andes." *Zoological Series of the Field Museum of Natural History* 24, no. 28:325-329.
Schoener, Thomas W.
1977 "Competition and the Niche." In *Biology of the Reptilia*, vol. 7, edited by Carl Gans and Donald W. Tinkle. London: Academic Press.
Seymour, Rev. William Wood
1898 *The Cross in Tradition, History, and Art.* New York and London: G. P. Putnam's Sons.
Sharon, Douglas
1978 *The Wizard of the Four Winds.* New York: Free Press.
Sherbondy, Jeanette
1979 "Les Réseaux d'irrigation dans la géographie politique de Cuzco." *Journal de la Société des Américanistes* 66:45-66.
Silverblatt, Irene
1976 "La organización femenina en el Tawantinsuyu." *Revista del Museo Nacional* (Lima) Vol. 42.
Skutch, Alexander F.
1976 *Parent Birds and their Young.* Austin: University of Texas Press.
Stein Callenfels, P. V. van
1931 "Bladvulling: Nog Enkele oud-Javaansche Sterrenbeelden terecht gebracht." *Tijdschrift voor Indische Taal-, Land- en Volkenkunde:*

Bataviaasch Genootschap 71:629-694.

Sullivan, William F.
1979 "Quechua Star Names." Dissertation submitted for the M.Litt. degree, University of St. Andrews.

Tastevin, P. C.
1925 "La Légende de Bôyusû en Amazonie." *Revue d'Ethnographie et des Traditions Populares* 6:172-206.

Tejeíro, Antonio
1955 "Nociones de una astronomía Aymara." *Khana* 3, nos. 11-12, 68-73.

Thom, A.
1967 *Megalithic Sites in Britain.* London: Oxford University Press.
1971 *Megalithic Lunar Observatories.* London: Oxford University Press.

Thompson, Donald E., and John V. Murra
1966 "The Inca Bridges in the Huanuco Region." *American Antiquity* 31:632-639.

Thompson, Philip D.
1961 *Numerical Weather Analysis and Prediction.* New York: Macmillan.

Traylor, Melvin A.
1952 "Notes on Birds from the Marcapata Valley, Cuzco, Peru." In *Zoology*, Chicago Natural History Museum Fieldiana series, 34, no. 3:17-23.

Tuan, Yi-Fu
1974 *Topophilia: A Study of Environmental Perception, Attitudes, and Values.* Englewood Cliffs, N.J.: Prentice-Hall.

Tuero Villa, Juan V.
1973 "Algunos rasgos tradicionales en la agricultura de la comunidad de Amaru." Thesis, Universidad Nacional San Antonio Abad del Cuzco.

Turton, David, and Clive Ruggles
1978 "Agreeing to Disagree: The Measurement of Duration in a Southwestern Ethiopian Community." *Current Anthropology* 19, no. 3: 585-600.

Urbano, Henrique-Osvaldo
1974 "La representación andina del tiempo y del espacio en la fiesta." *Allpanchis Phuturinqa* 7:9-48.
1976 "Lenguaje y gesto ritual en el sur andino." *Allpanchis Phuturinqa* 9:121-150.
n.d. "Representación memoria en los Andes." Ms.

Urteaga, Horacio H.
1913 "Observatorios astronómicos de los Incas." *Boletín de la Sociedad Geográfica de Lima* 29:40-46.

Urton, Gary
1978a "Beasts and Geometry: Some Constellations of the Peruvian Quechuas." *Anthropos* 73:32-40.

1978b "Orientation in Quechua and Incaic Astronomy." *Ethnology* 17, no. 2:157-167.

1979 "The Astronomical System of a Community in the Peruvian Andes." Ph.D. dissertation, University of Illinois, Champaign-Urbana.

1980 "Celestial Crosses: The Cruciform in Quechua Astronomy. *Journal of Latin American Lore* 6, no. 1:87-110.

1981 "Animals and Astronomy in the Quechua Universe." *Proceedings of the American Philosophical Society,* vol. 125, no. 2:110-127. no. 1, in press.

Valcárcel, Luis E.

1943 "Leyenda del arco-iris: Las dos serpientes." *Folklore* 2:99.

1963 "The Andean Calendar." *Handbook of South American Indians,* 2:471-476. Washington, D.C.: Bureau of American Ethnology.

Vallée, Lionel

1972 "Cycle écologique et cycle rituel: Le cas d'un village andin." *Canadian Review of Sociology and Anthropology* 9, no. 3:238-254.

Vargas C., César

1936 "El Solanum Tuberosum a través del desenvolvimiento de las actividades humanas." *Revista del Museo Nacional* (Lima) 5, no. 2:193-248.

von Humboldt, Alexander

1975 *Views of Nature* (1850). New York: Arno Press.

Wachtel, Nathan

1971 *La Visión des vaincus.* Paris: Editions Gallimard.

Wagner, Catherine A.

1976 "Coca y estructura cultural en los Andes peruanos." *Allpanchis Phuturinqa* 9:193-223.

1978 "Coca, Chicha and Trago: Private and Communal Rituals in a Peruvian Community." Ph.D. dissertation, University of Illinois, Urbana-Champaign.

Walker, Ernest P., [and others]

1964 *Mammals of the World.* Baltimore: The Johns Hopkins Press.

Wallis, Christopher N.

n.d. "Dependence and Independence among a Group of Llama-herders in the Highlands of Southern Peru." Ms.

Weckmann, Luis

1951 "The Middle Ages in the Conquest of America." *Speculum* 26: 130-141.

Weiss, Gerald

1969 "The Cosmology of the Campa Indians of Eastern Peru." Ph.D. dissertation, University of Michigan, Ann Arbor.

Wheatley, Paul

1971 *The Pivot of the Four Quarters.* Chicago: Aldine-Atherton.

Wilbert, Johannes

1975 "Eschatology in a Participatory Universe: Destinies of the Soul among the Warao Indians of Venezuela." In *Dumbarton Oaks Conference on Death and the Afterlife in Pre-Columbian America,* edited by E. P. Benson, pp. 163-190. Washington, D.C.: Dumbarton Oaks Research Library and Collections.

Zuidema, R. Tom

1964 *The Ceque System of Cuzco.* Leiden: E. J. Brill.

1966 "El calendario inca." *36° Congreso Internacional de Americanistas, Actas y Memorias* (Seville) 2:24-30.

1968 "Un modelo incaico para el estudio del arte y de la arquitectura prehispánicas del Perú." *Verhandlungen des XXXVIII Internationalen Amerikanistenkongresses* 4:67-71.

1973 "Kinship and Ancestorcult in Three Peruvian Communities: Hernández Príncipe's Account in 1622." *Bulletin Institut Français des Études Andines* 2, no. 1:16-33.

1977a "The Inca Calendar." In *Native American Astronomy*, edited by Anthony F. Aveni, pp. 219-259. Austin: University of Texas Press.

1977b "The Inca Kinship System: A New Theoretical View." In *Andean Kinship and Marriage,* edited by Ralph Bolton and Enrique Mayer, pp. 240-281. American Anthropological Association, Special Publication, no. 7. Washington, D.C.

1977c "Mito e historia en el antiguo Perú." *Allpanchis Phuturinqa* 10: 15-52.

1978a "Lieux sacrés et irrigation: Tradition historique, mythes et rituels au Cuzco." *Annales*, nos. 5-6:1037-1056.

1978b "Mito, rito, calendario y geografía en el antiguo Perú." *Actes du XLIIe Congrès International des Américanistes* (Paris) 4:347-357.

1980 "El Ushnu." *Revista de Universidad Complutense* (Madrid), vol. 2, no. 117:317-362.

n.d.a "The Labyrinth and Straight Line as Ritual of Form: Problems in the Prehistory of Science." Ms.

n.d.b "Professor John Howland Rowe and the Study of Inca Astronomy." *Latin American Research Review,* in press.

n.d.c "The Inca Observation in Cuzco of the Solar and Lunar Passages through Zenith and Anti-Zenith." In *Archaeoastronomy in the Americas,* edited by Ray Williamson. Santa Barbara, Cal.: Ballena Press, forthcoming.

Zuidema, R. Tom, and Gary Urton

1976 "La constelación de la Llama en los Andes peruanos." *Allpanchis Phuturinqa* 9:59-119.

Index

Acllas, 79
Acosta, José de, 4, 6
Adhara, 103, 179
Agriculture: in the calendar of Misminay, 25; in the daily cycle, 19; in the weekly cycle, 22; llama sacrifices in Inca, 187; lunar phases related to, 85; rain harmful to, 176; relation between celestial phenomena and, 193, 194; role of Pleiades in, 118-121
Aguilar Páez, Rafael, 87
Albó, Javier, 32, 42
Albornoz, Cristóbal de, 178
Allen, Catherine, 26, 51, 67, 176. *See also* Wagner, Catherine Allen
Alpaca, 187
Alpha Crucis: center of Milky Way; 161; heliacal rise and set of, 184; point of union of celestial rivers, 202; relation to south celestial pole, 59; rising point of, from Cuzco, 201
Alqo (constellation), 208
Altair, 214
Altomisayuq (priest/diviner), 92
Alto piña ch'aska (midnight zenith star), 96
Alto qana (constellation), 208
Amaru: as rainbow serpent, 88-89, 177-179. *See also* Serpent
Amaru (District of Pisac), 125-126
Amaru Cocha (lake), 111
Amaru Contor (constellation), 97
Ancascocha: direction to, 71
Ancestors, 48, 51-52
Androgyny, 89, 109
Animals: animal-marking ceremonies, 129-130; dark cloud constellations, 109-110, 169-191; origin of, 93; procreation of, 189; reproduction of, 196
Anonymous (1906), 6, 77, 196, 197, 199
Antisolstice, 70, 116
Antizenith, 76, 188, 197
Apu. *See* Sacred Mountains
Arado (constellation), 98, 207, 209. *See also* Plow
Aranguren Paz, Angélica, 175
Archaeoastronomy, 211
Arguedas, José María, 43
Aristotle, 198; theory of Milky Way, 214-215
Arriaga, Pablo José de, 4, 77, 132, 200
Arroyo Medina, Patricio, xviii, 13, 32, 212

Atoq (constellation), 102, 170-171, 188-189, 208. *See also* Fox
August: crop predictions, 11, 118-122, 124-125, 173-174; divination for rain in, 181; heliacal rise of dark cloud Serpent, 179; planting in Inca Cuzco, 77, 197; planting in Misminay, 74, 76; weaving month, 30
Ausangate (apu), 116, 175
Aveni, Anthony F., 6, 7, 76, 130, 159, 213
Avila, Francisco de, 9, 119, 170, 188, 200
Ayllus: and constellations in Recuay, 10, 134; in Maras, 13
Ayni (reciprocal labor), 25, 31

Baity, Elizabeth C., 211
Bárcena Cruz, Carlos E., 212
Barrette, Christian, 47, 193
Barthel, T. S., 120
Bastien, Joseph W., 193
Beans, 25, 64
Becher, Hans, 85
Beetle, rhinoceros, 97
Berry, Arthur, 198
Betanzos, Juan Diez de, 6
Betelgeuse, 98, 99
Bilingualism: in Misminay, 13, 20
Biological metaphor: for lunar phases, 80, 82, 84
Biological standard: for lunar zodiac, 79
Blood, 85
Boa, 178
Boca del Sapo (constellation), 98
Boleadora, 92-93, 96. *See also* Shooting stars
Bolton, Ralph, 19, 26
Bonett Yépez, Jorge W., 11, 125-126, 207, 214
Boulenger, George A., 177
Boyer, Carl B., 87
Bóyusú (constellation), 178
Bridge (constellation), 99, 131-132, 139. *See also Chaka*
Brownrigg, Leslie Ann, 27, 74
Brush, Stephen, 13, 25
Bufo spinulosus, 180
Bull: black, associated with rain, 176; constellation of little, 97; pasturing of, 18
Burro, 18, 38

Cabañuelas (constellation), 98
Cabrillas (constellation), 113, 119. *See also* Pleiades
Cadogan, León, 178
Calca (Department of Cuzco): direction to, 71; in the solar calendar of Misminay, 74, 76; observation of sunrise from, in August, 197; as subhorizon city from Misminay, 51
Calendar: of agriculture in Misminay, 24, 25; Andean, 12-13; correlation of activities and celestial phenomena in, 195; correlation of synodic and sidereal lunar, 79; horizon and Misminay calendar, 74; Incaic and contemporary, 197; llamas in the Inca, 187; moon in the Inca, 74; of saints' days and solar observations, 77; of toads and agriculture, 181-182; of tinamous and agricultural guard duties, 185-186; saints' days, 27-29
Callegari, G. V., 5
Calvario (cross of the axes of the Milky Way), 55, 64
Calvario Cruz (constellation), 98, 134-136, 140-147
Canchis, Province of, 64
Canis Major, 99, 179
Canopus, 214
Capac Raymi, 90
Carnaval, 129
Carrión Cachot, Rebeca, 137
Casaverde Rojas, Juvenal, 11, 67, 68,

81, 82, 88, 90, 175, 193, 207, 213
Castor, 98, 99
Cat: golden, 99; *k'owa*, the black, 90; lightning as a, 92; sacred mountain of, 51
Catachillay (Inca constellation), 130-131
Catalogue of Quechua Stars and Constellations, 96-105
Cei, José M., 180
Celestial pole: dark cloud constellations and the south, 170; inclination of the Milky Way from, 163, 201; and Milky Way in Chinese cosmology, 212
Centaurus, 100, 101, 103, 104, 187, 208, 209
Center sun: azimuths of, 75; concept of, 72; in the horizon calendar of Misminay, 75-77. *See also* Sections of the sun
Ceques: and Andean pilgrimages, 211; antizenith orientation of, 76; definition of, 8; "line", xvii; orientation of the *suyus* of the, 201
Chaka (Chaca): in Quechua cosmology, 150; term for "cross", 130-133; uses of concept in Misminay, 138-140. *See also* Bridge; Cross
Chakana, 99, 200
Chakma, 26, 27
Ch'aska, 107, 207. *See also* Stars
Ch'aska plata, 96. *See also* Shooting stars
Chavín; 199
Chicha: libations of, 130; preparation of, 139; times of consumption of, 19, 20, 26, 31
Chillkapampas (District of Maras), 13
Chimpu. *See* Halo
Chinchaya, 90. *See also* *K'owa*
Chinchero (Province of Urubamba): and center sun, 75; celestial cross in, 132; direction to, from Misminay, 71; rise of sun from, 72, 76; solar terminology in, 67
Ch'issin ch'aska, 20, 96, 151, 156, 167, 200. *See also* Evening star
Choquechinchay: Quechua constellation, 99, 114; Inca constellation, 200
Chumbivilcas, Province of: cross terminology in, 135
Ch'uño, 51
Chuqui (constellation), 132
Clock: Milky Way as, 55; sun moves as, 68
Coalsack: in Chumbivilcas, 134; in dark cloud constellations, 102, 103, 104, 181, 207, 208; relation to south celestial pole, 59; rising point of, in Cuzco, 201. *See also* *Yutu*
Cobo, Bernabé, 4, 6, 8, 180, 183, 188, 196, 200, 202, 204
Coca: 19, 55
Cocha (lake/reservoir), 43
Colla Capac, 80-81, 197. *See also* *Coya Capac*; Moon
Collana *ayllu*, 13
Collca: axis of, 116-117; constellation of Storehouse, 3, 98, 113-127, 207, 213; Inca constellation, 200; "mobile" star of, 166
Color: light and dark classifications, 109-111; multicolored eggs of the tinamou, 184; multicolored serpents, 178; of clothing stolen by rainbow, 90; of fire, sun, and *sullaje*, 92; of haloes, 69, 91; of llamas sacrificed in Inca calendar, 187; of rainbows, 87-88
Colparay (District of Maras), 40, 51
Compadrazgo, 13, 67
Condores: beings in underworld (*otra nación*), 38, 212

Condori, Bernabé, 175
Contor: vulture constellation, 99, 102, 136, 208; distinction between *contor* and *condores*, 212; Inca constellation, 200; on Wañumarka, 70
Coordinate system, 45, 52
Copernicus, N., 198
Corimaylla: ancestor of Misminay, 48
Corona Borealis, 100, 102
Corpus Christi (festival), 68
Coscotoca ch'aska (midnight zenith star), 96, 158
Cosmic mountain: *Volcán*, 37
Cosmic river, 69, 202
Cosmic sea, 37, 59, 68, 172, 176
Coya Capac, 80, 197. *See also* *Colla Capac*; Moon
Coyllur, 107. *See also* Star
Coyllur Riti (festival), 68
Crop predictions: celestial observations for, 11, 118-121, 193, 196, 207, 209; Inca celestial observations for, 200; in the community of Amaru, 125-126; made on San Juan, 23. *See also* August
Cross: cesestial, 98, 99, 129-150, 207, 208; constellations, 55; intersection of terrestrial axes, 52; intersection of terrestrial and celestial axes, 64; of axes of the Milky Way, 61; of Calvary, 138; of zenith, 65; on horizon, 51; origin of term *linun*, 214; paired, 140-143; symbolism of, 147-150; syncretism of Catholic and Inca concepts of, 68; *taytacha* as term for, 67; terminology of stellar, 130; terminology and identification of stellar, 134-147
Crossbeam: 132, 139
Crucero: center of terrestrial axes, 64, 65; chapel in Misminay, 18, 52; crossing of footpaths at, 40; crossing of hydraulic axes at, 44; house cluster of, 46; view of horizon from, 48
Cruz Calvario (constellation), 134-136, 208, 209; intersection of Milky Way at zenith, 61. *See also* Cross
Cruzero Estrellas (Inca constellation), 130-131. *See also* Cross
Cuscan: related to lunar phases, 84, 157, 158. *See also* Synodic Lunar Cycle
Cuzco: altitude, climate, 12; direction from Misminay to, 71; in Inca cosmology, 201-202; local calendars in area of, 194; quadripartition of Inca, 42, 201; solar pillars in, 6-8, 77

Daily cycle, 17-21. *See also* Week
Dalle, Luis, 26
Dark cloud constellations: catalogue of, 102-104; definition of, 109-111; description and discussion of, 5, 10, 11, 169-191, 207, 208, 209; recognition of, 95; related to spots of moon, 81
Dawn, 152-153. *See also* Morning star
Dawn of the earth/time star. *See* *Pachapacariq ch'aska*
December solstice: foxes and, 70, 189; and rainy season, 62; related to dark cloud constellations, 172, 184-185; related to drums, 90; related to Milky Milky Way, 161-165; in Sonqo, 116
Demon: black cat, 90; devil, 51, 180; lightning as, 92
De Schauensee, Rodolphe M., 183
Directions, cardinal: 5, 55; cosmological, 59, 71; term for east, 68; term for west, 68. *See also* Intercardinal directions

Ditmars, Raymond L., 178
Divination: by behavior of toads, 181, 200; by coca and maize, 23; by shamans in Urubamba Valley, 177; by shooting stars, 87, 92; by shooting stars and Milky Way, 64; by stellar scintillation, 93
Division of labor: by age, 20; by sex, 18, 19, 20, 22
Double-headed creatures: *amarus,* 178; serpents, 89, 178
Doughty, Paul L., 26
Drums: Incaic symbolism of, 90; related to the seasons, 30-31; type of rainbow, 89
DuGourcq, Jean, 5
Dusicyon culpaeus, 188
Dusk, 153. *See also* Evening star; *Ch'issin ch'aska*
Duviols, Pierre, 175, 178, 201

Earls, John, xvii, xviii, xix, 10, 90, 116, 150, 193
Earth: related to dark cloud constellations, 174-177; shape of, 37, 59, 63. *See also Pachatira*
Ecliptic, 9, 70, 121, 141, 147
Eggs, tinamou, 184
Equator, 121, 139, 141
Equinox, 71, 74, 139, 172
Escobar, Gabriel, 119
Ethnoastronomy, 211
Evening star, 95, 107, 156, 197, 207; *See also Ch'issin ch'aska*
Ewer, R. F., 189

Fallowing cycle: crop rotation, 12, 30, 32-35
Father: name applied to stars, 109, 134; *taytay*, 107; *yaya*, 107
February: crop predictions, 11; heliacal set of dark cloud serpent, 179
Fecundity: of earth and dark cloud constellations, 175-176, 189
Ferguson, George W., 138
Fertility: of water and semen, 202; water as source of, 93, 172, 189
Fertilizer, 30
Fioravanti, Antoinette, 12
Fireplace, 176
Fitch, Henry S., 177
Flute: *quena*, 30-31
Foam, 59, 111, 202-204ç
Fock, Neils, 43, 63
Foetus, llama, 26
Fonseca Martel, César, 42
Footpaths: and division of community space in Misminay, 40-43, 46; associated with irrigation canals, 53, 129, 195
Foot-plow, 27
Forked cross, 140, 143. *See also* Cross
Foster, George M., 70
Fox: axis of the December solstice and the, 70, 71, 116, 190; dark cloud constellation of, 188-189; mating season of, 189
Franklin, William L., 188
Friedmann, Herbert, 185
Frogs, 180
Frost, 27, 76, 181, 208
Fulcrum, 43

Gade, Daniel W., 177
Gaetano de Thiene, 81
Gaposchkin, Sergei, 11
Garcilaso de la Vega, El Inca, 6, 169, 177
Gemini: in Chinese astronomy, 212; in Quechua astronomy, 99
Gomarra-Thompson, Rosa, 21
González Holguín, Diego, 9, 81, 130, 131, 132, 156-158, 200
Gow, David D., 19, 67, 74, 175, 193, 212, 213
Gow, Rosalind, 175

Graham, F. Lanier, 88
Grzimek, Bernard, 180, 183, 184
Guamán Poma de Ayala, Felipe, 74, 185, 197, 200
Guardia Mayorga, César A., 130-131

Hagar, Stansbury, 5
Hair: copper, 214; of rainbow, 90; stars called shaggy, 107; time for cutting, 67
Hallmiyoq, 26
Halo: in Quechua meteorology, 90-91; prediction of rain by, 213; solar, 69.
Hanan moiety, 42
Hanp'ătu (constellation), 102, 105, 170-171, 180-181, 208. *See also* Toad
Harvest: and pasturing cycle, 30; consumption of alcohol during, 26; determining the time of, 85; in agricultural calendar, 27; music associated with 31-32; and pasturing cycles, 30; predicting size of, 120; sunrise at time of, 72, 74
Hatun Coyllur (star), 96
Hatun Cruz (constellation), 99, 134-135, 140-147, 207. *See also* Cross
Hatun Llamaytoq (constellation), 103
Hawkins, Gerald, 115
Heaven, 63
Heliacal, 159, 214; definition of, 213
Hemispheres, 57, 61, 139, 141, 157
Hibernation: snake, 179; toad, 180
Hierarchy, 46-48
Hoeing: *lampara*, 26
Holq'e (constellation), 208
Horizon: calendar, 72, 73; celestial and terrestrial quadripartitions meet at, 116; intersection of, by terrestrial and solstitial axes, 64-65; landmarks on, 54, 116; places of rise and set of Milky Way along, 161-165; space traversed by sun, 71
Horse: in spots of moon, 81
House clusters, 45-48, 65
Huacas: definition of, xvii; destruction of, 4; in Cuzco, 8; snakes as, 178
Huancaraylla (province of Víctor Fajardo), 47
Huaylanka: and center sun, 74, 76; related to foxes and solstice, 70; set of sun at, 72
Huayna Capac (Inca King): living in Moray, xviii, 51; as name for sun, 67, 68, 80, 81, 196
Huchuy Cruz (constellation), 99, 134-135, 141-147, 208. *See also* Cross
Hudson, W. H., 183
Hurin moiety, 42, 47
Hyacinth, 180
Hyades, 5, 98, 101
Hydra, 99

Illapa. See Lightning
Illarimi ch'aska: 97, 151, 156, 167, 200. *See also* Morning star
Inca: as ancestors, 48; astronomy compared to contemporary Quechua astronomy, 196-204; astronomy in general, 6-10; dark cloud constellations, 169-170, 176, 177; Huayna Capac, 51; initiation of nobles, 187-188; kings in solar terminology, 67-68; llamas in the calendar of, 187; planting festivals and solar observations, 77; pottery, xviii; stars and constellations, 200; terms for crosses, 138; times for

guarding crops in calendar of, 185-186; Virgins of the Sun, 79
Intercardinal Directions: axes, 64; in celestial orientations, 65; cosmological, 61, 62, 197; of Collca axis, 116; the "directional cross", 136; in Inca cosmology, 201, in Juncal, 45; in organization of terrestrial space, 40-48; and Milky Way, 57, 59, 68, 94, 165; and rainbows, 94; and the solstices, 70, 94; of Vilcanota River, 56, 68
Inti. See Sun
Irrigation, 31, 57, 64
Irrigation canals: association of reservoirs and, 53; cleaning of, 68, 138, 202; and footpaths, 195; in Huancaraylla, 47; and Milky Way, 69; in Misminay, 43-45; orientation of, 64; in Quechua cosmology, 56-57; and the *wamanis*, 137
Isbell, Billie Jean, 13, 19, 45, 74, 129, 132, 137, 138, 193, 202
Ismalia (star), 208

Jaki, Stanley L., 81, 198, 215
Jara, Victoria de la, 4
Jawa moiety, 45
Jesus Christ: and sacred mountains, 51; name for the sun, 67, 80
Juncal (province of Cañar, Ecuador), 45, 63
June solstice: and festival of San Juan, 23; Inca pilgrimage on, 201-202; orientation of Milky Way at, 62, 161-165; and planting moon, 123-124; prognostications made on, 118-119, 121-122; and sacred mountains in Sonqo, 116; and tinamou constellation, 184-185

Kaqllarakay (District of Maras), 40
Kelly, David H., 9
Kepler, John, 214
Khaswa Coyllur (constellation), 100
Kinship, 54
Kkoto. See Qutu
Knowlton, Frank H., 183
K'olli (constellation), 103
K'otin (sacred mountain), 51, 115. *See also* Sacred Mountains
K'owa: black cat, 90; related to lightning, 92
Kulkarni, B. R., 212
K'uychi. See Rainbows
Kuyo Chico (Province of Calca), 20
Kuyo Grande (Province of Calca), 67, 81

Lagercrantz, Sture, 109
Lancaster, D. A., 183
La Raya: birth of sun at, 201
Lares, 13
Larrea, Juan, 92
Lathrap, Donald W., 199
Leach, Edmund, 7
Lehmann-Nitsche, Robert, 9-10, 132, 170
Lévi-Strauss, Claude, 85, 109, 191
Lewy, Hildegard, 213
Light cloud constellations: catalogue of, 105; discussion of, 111
Lightning, 91-92
Linun Cruz (constellation), 100, 135, 141-147; origin of name, 214. *See also* Cross
Lipkind, William, 59
Lira, Jorge, 11, 42, 87, 89, 91, 113, 118, 173, 178, 193, 207
Llama: in community of Amaru, 126; in relation to *Calvario*, 134, 136; foetus of, 26; related to fox, 188; in Inca astronomy, 10,

169-171; multicolored, 187-188; in Quechua astronomy, 103, 185-188, 208, 209; in spots of the moon, 81
Llama Cancha (constellation), 100
Llamacñawin (constellation), 100, 187, 200, 207, 208
Lloque Yupanqui (Inca king), 90
Locero (star or planet), 97, 156. *See also* Morning star
Lucre (District of Oropesa): cross terminology in, 134, 135; location and description of, 212
Lumbreras, Luis, xviii

Mach'ácuay (constellation of serpent), 88, 100, 103, 170-171, 200. *See also* Serpent
Maegraith, B. G., 109
Magellanic cloud: large and small, 208
Maize: germinating kernels, 26; growing season, 27-29; planting of, 25, 85; relation to Pleiades, 118, 120-121; relation to Sirius, 118; solar observations for planting, 76-77
Mama Rosario (constellation), 105
Mamacha Asunta (festival), 29
Mamacha de las Mercedes (festival), 29
Mamana Micuc (constellation), 100, 200
Manco Capac (Inca King): associated with sun, 68, 81, 196
Manta, 202
Mar, 68. *See also* Cosmic sea
Maras (province of Urubamba), 13, 20, 21, 32, 48, 212
Mariscotti de Görlitz, Ana María, 87, 213
Markets, 21-22
Mayer, Enrique, 193
Mayu. *See* Milky Way
Menstrual cycle, 79, 85, 198
Midnight, 107, 197; zenith star, 96, 97, 107, 153-156, 166-167
Milky Way, 103, 207, 208; animals and water in, 172-174; Aristotle on, 198; bridges crossing, 139; in celestial orientations, 54-55; in Chinese and Indian cosmology, 212; constellations in, 10, 108-111, 169-171; constellations of, in Amaru 125-126; cosmic circulation of water 69; cross of axes of, 136, 196; in Inca cosmology, 204; Inca orientation based on, 201-202; orientation by, 9; passage of sun through, 70; as plane of orientation, 147, 195; quartering of, 159-166; in Quechua cosmology, 56-65, 197; and rainbows, 93-94; rainbows, the sun, and 190; relationship of sun and foxes in, 189; and the Vilcanota River, 38
Minnaert, M., 89, 94
Mirco Mamana (constellation), 200
Mirror: mirror image of celestial and terrestrial quadripartitions, 63, 150, Vilcanota River as, reflecting the Milky Way, 56
Misa, 130
Misa Tupac Amaru: ancestor of Misminay, 48
Mishi (constellation), 208
Mishkin, Bernard, 11, 68, 81, 90, 119, 193, 194, 207
Misminay: definition of name, xvii; description of community space, 40-48; location and description of, 13
Mistirakai, xvii-xix, 13, 70
Mitchell, William P., 27, 31, 43, 45
Moieties, 42, 46
Molina, Cristóbal de, 201
Monocerotis, 99
Months: duties of the, 23-29; May-August, 62; November-February,

62; relation to sunrise points, 74
Moon: and agriculture, 27, 195; and celestial fox, 70; cosmology and astronomy of, 77-85; crops and animals related to, 194; as a female, 74; full, and antizenith sun, 76; half, 42; haloes around, 91; in Inca calendar, 74; in Inca cosmology, 197; in Inca studies, 199; and months, 23; planting, 122-125; terms for quarters of, 156-158
Moray: and Apu Saqro, 52; description, xvii, xviii; Incas now living in, 48; "internal place", 42; pasturing in, 30; people living in area of, 13, 15; as sacred mountain, 51; stories about, xviii; in studies of Inca cosmology, 10
Morissette, Jacques, 42, 193
Morning star, 95, 97, 107, 156, 197, 207. *See also Illarimi ch'aska; Pachapacariq ch'aska*
Morote Best, Efraín, 23, 26, 116
Mother Moon, 80-81
Mountford, Charles P., 109
Mullaca, xix, 13
Müller, Rolf, 6
Murra, John V., 13, 17, 34, 187
Musical instruments, 30-32

Nachitgall, Horst, 26
Nadir. *See* Antizenith
Ñan, 209
Navidad (festival), 70, 197
Ñawin Cristo (constellation), 101
New Year's: and birth of foxes, 70; in Sarhua, 116
Nilsson, Martin P., 214
Nimuendajú, Curt, 109
Noble, G. Kingsley, 180
Northern Cross: Cygnus, 147
Number: of colors in rainbow, 87-88; of stars in Pleiades, 113; terms to state time of day, 68
Núñez del Prado, Juan Víctor, 11, 21, 67, 74, 81, 82, 89, 175
Núñez del Prado, Oscar, 20, 21, 51, 81, 175
Nussenzveig, H. Moyses, 88
Ñust'as (princesses), xviii

Oca, 25
Ocean, 63. *See also* Cosmic sea
Ochoa G., Luis, 132
October: end of planting period, 76; rise of celestial toad, 181
Oesch, Will A., 87, 89
Ombligo de la Llama (constellation), 104, 200
Oncoy (Inca constellation), 200
O'Phelan Godoy, Scarlett, 13
Opposition: of celestial crosses, 141-142; of color, 189; of Collcas, 114; of dark and multicolored serpents, 178; of intercardinal axes, 195; of Pleiades and Scorpio, 118; of solstices and crosses, 143-147; of sun, rainbows, and the Milky Way, 190
Orcochillay: Inca constellation, 130-131; Urquchillay, 200
Orientation: of celestial and terrestrial river, 57; of Collca axis, 116-117; for the divisions of space, 195; of local rivers and irrigation canals, 64; by Milky Way, 59-61, 62, 201-202; of Milky Way, 9; of Milky Way and rainbows, 93-94; of the *pachapacariq ch'askas,* 159-160; of space by Collca axes, 118; in terrestrial space, 45; in viewing night sky, 54
Origin: of animals, 93; of animals in Milky Way, 174-176; of dark and light cloud constellations, 111; of Milky Way, 59-60; of

Misminay, 47-48; quarter of, 64; related to the north, 51; of stars, 151; of the universe, 201; of water and people, 47
Orion, 5, 98, 99, 101, 107, 130, 131, 132, 134, 138-139, 207, 208
Orlove, Benjamin, 23, 64, 70, 118-119, 207, 208
Ortiz Rescaniere, Alejandro, 111, 176
Ossio, Juan, 43, 204
Otra nación (underworld), 38, 63, 68, 151, 212
Otuño: name for sun, 67
Oveja (constellation), 208
Oyola, 13

Pachacuti Yamqui Salcamaygua, Juan, 5, 9-10, 87-88, 90, 114, 132, 200, 202, 215
Pachakilla: name for moon, 81
Pachakuti: toad, 180
Pachamama: earth mother, 73, 81, 93, 175
Pachapacariq ch'aska: dawn of earth/time star, 21, 97, 122, 151, 156, 158-167, 200, 207, 214. *See also* Morning star
Pachatira: as classification of dark cloud constellations, 109, 174-177; fecundity of earth, 93, 189
Pachawawa: toad, 180
Palm tree, 38
Palomino Flores, Salvador, 42, 129, 130
Pampakonas, 13
Pannekoek, A., 109
Papa Dios Cruz (constellation), 101, 135. *See also* Cross
Papa pachapacariq ch'aska (star), 107
Paqo: diviner, 19, 80, 92; constellation (*paqocha*), 208
Passon Cruz (constellation), 101, 134, 135. *See also* Cross
Pastoralism: and fallowing, 31-32; and lightning, 91-92; and the moon, 81; in Misminay, 18; at night, 21; and the origin of Misminay, 48; seasonal cycle of, 30, 197; weekly cycle of, 21, 22
Paucartambo, 51
Pauchis, 13
Paz Flores, Percy, 82, 208
Pearson, A. K., 185
Pease, Franklin, 202
Pichiko (bird): call of, 17, 18, 21, 151
Pichinqotos, 13, 25, 31
Pikillaqta, xviii, 212
Pilgrimage: and the ceque system, 211; Inca, on June solstice, 201
Pillars: for solar observations in Inca Cuzco, 6, 8; saints' days and solar, 77, 197
Pinto Ramos, E., 202-203
Pisqa Collca (constellation), 101-200
Pisqa Coyllur (constellation), 101, 200
Pitusiray, 116
Planets: crop prediction by observing, 106-107, 166-167, 199; Jupiter, 122; the Pleiades and, 121-122. *See also* Venus
Planting: cycle in Misminay, 25-29; determined by sun, 72-73; in the Inca calendar, 196; moon and the Collcas, 122-125; time of, by observing Pleiades, 118-121; timed by sunrise from Calca, 74
Pleiades: celestial Storehouse (Collca), 98, 101, 113-127, 207, 208, 209; cycles of maize and, 120-121; Inca crop predictions by observing, 200; rise of, from Misminay, 115; and the solstitial axes, 71

Plow, 27; constellations of, 125-126. *See also Arado*
Pocock, D. F., 211
Polaris, 54
Polo de Ondegardo, Juan, 6, 8, 9, 132, 169, 176, 179, 187, 191, 200
Poole, Deborah A., 211
P'oqroy, 26
Posener, Georges, 211
Posuqu, 59, 105, 111, 202-203
Potatoes: field names, 33; harvest of, 27; lunar phase for planting, 85; planting of, 25; prognostications by selecting, 119; rotation of, 32-34
Procyon, 98, 99
Prognostications: by observing dark clouds, 173-174; by observing June solstice, 118-119. *See also* Crop predictions; Divination
Ptolemy, 198
Pukaras, 13
Pukiu (spring), 88-89
Puma (constellation), 208
Puppis, 100
Pura: phases of moon, 82-85, 157
Pyxidis, 99

Qasa (constellation), 208
Q'enco Mayu, 51
Qholla: term for moon, 80-81
Qoripujios, 13
Quadripartition: of celestial sphere by lunar phases, 157; by Collca and Atoq axes, 116-117; of community space, 195; of Huancaraylla, 47; Milky Way quarters, 159-166; of Misminay, 42-45; of sky, 55-65, 143-147; symbolism of, in Misminay, 53-54. *See also Suyus*
Quinua, 25
Quipu, 8
Quiroga, Adán, 130, 138
Quisqamoko (sacred mountain), 51
Quispihuara (District of Santa Ana), 212, 213
Qutu: constellation, 101, 207, 208, 209; crop predictions by observing, 121; Pleiades, 113. *See also* Tinamou

Racine, Luc, 42, 193
Radicati di Primeglio, Carlos, 4
Rain: in cycling of water, 60; harmful to agriculture, 176; motive of American Indian religions, 138; prediction of, 69, 91, 173-174, 181
Rainbows: classification of, 213; in cosmology, 93-94; and dark cloud constellations, 178-179, 185; in Quechua meteorology, 87-90; mediators of subterranean and human water, 176; relationship with Milky Way, 190; and tinamou eggs, 184
Raki, 42, 139-140
Ravines Sánchez, Roger, 71
Rayo. See Lightning
Rees, William E., 183
Reichel-Dolmatoff, Gerardo, 85, 109, 147, 184
Reijners, Gerardus Q., 214
Relámpago. See Lightning
Reservoirs, 42, 43-45, 46, 53, 64
Rigel, 98, 99
River, 55, 56, 60, 62, 64, 111, 188. *See also* Cosmic river; Milky Way
Roca W., Demetrio, 180-181
Rodríguez Rivera, Virginia, 87
Roe, Nicholas A., 183
Rowe, John H., 7, 199
Ruggles, Clive, 7

Sacred Mountains: equated with crosses, 129, 137; in the organization

of terrestrial space, 48-53, 54
Sagittarius, 70, 102, 189; in Chinese astronomy, 212
Saint: sun, 67
Saints' days: and horizon calendars, 197; and planting of maize, 118-119; and solar observations, 67, 76-77
Sallnow, Michael J., 68
Samanez Argumedo, R., xviii
San Francisco (festival), 29
San Juan (festival): beginning of Misminay calendar, 23; crop predictions made on, 118, 121-122; related to solstices, 70, 197
Santa Ana (Province of Urubamba), 13, 20, 40, 129
Santa Ana (festival), 29
Santillana, Giorgio de, 212, 213
Santissima Cruz (constellation), 135. *See also* Cross
Sapo (constellation), 208. *See also* Toad
Saqra, 180
Saqro (sacred mountain), 51
Sarmiento de Gamboa, Pedro, 6, 35
Schafer, Edward H., 212
Scharfe, Hartmut, 212
Schaumberger-Gars, J., 158
Schiller, Gertrud, 138
Schlegel, Gustaf, 212
Schmidt, Karl P., 177
Schoener, Thomas W., 179
School, 20
Scintillation, stellar, 87, 93
Scorpio: as Collca, 113; constellation, 97, 98, 99, 100, 101, 102, 103, 104, 105, 208, 209; cross constellation, 134, 136; and dark cloud fox, 189; and dark cloud serpent, 178; relationship of moon and June solstice to, 213; rise of, from Misminay, 115; tail of, 70, 71
Scutum, 104
Seasons: breeding of toads in rainy, 180; cycle of, related to celestial serpent, 179; cycling of, 195; determined by Milky Way, 165; different colored haloes related to, 91; divisions and activities of, 29-32; drums and rainbows related to rainy, 89; dry, 62; and hibernation of snakes, 179; Milky Way orientation during rainy, 71; rainy, 61; related to size of sun, 69; and visibility of dark cloud constellations, 172
Sections of the sun: related to Inca solar pillars, 196-197; the three, 71-74
Semen, 202
Señoracha (constellation), 208
Serpent: as rainbows, 88-89, 93; constellation, 176-180; related to seasons, 172
Sex: division of labor by, 18; of celestial crosses, 134; of dark cloud constellations, 109, 174-175; of earth, 73; of haloes, 91; of *K'owa*, 90; of light cloud constellations, 111; of lightning, 91-92; of moon, 79-81; of rainbows, 89; of stars, 107; of star-to-star constellations, 107, 109; of sun, 67, 80; of sacred mountains, 51-52, 53; school attendance by, 20
Seymour, William W., 138
Sharon, Douglas, 132
Shells, 31
Sherbondy, Jeanette, 43
Shooting stars: as cattle thief, 208; *ch'aska plata* and *boleadora*, 92-93; divination by, 64, 87; source of subterranean silver, 92
Sidereal lunar cycle: definition of, 77, 78, 79; in Inca calendar, 197; related to Milky Way, 165

Silverblatt, Irene, xviii, 10, 90, 150
Sirius, 96, 98, 99, 118, 167, 214
Skunk, 70
Skutch, Alexander F., 185
Sol. See Sun
Solstice: and celestial axes, 65, 190; and celestial crosses, 141-143; and centers of quarters of Milky Way, 161-165; discussed in general, 70-71; and foxes, 189; in the Inca calendar, 197; related to rainbows and Milky Way, 93-94; seasons and Milky Way related to, 62. *See also* December solstice; June solstice
Sonaja (constellation), 102
Sonqo (District of Qolqepata): celestial and terrestrial axes in, 116; constellation of *Sulluullucu* in, 178; cross terminology in, 135, 138, 140-141; location and description of, 12, 59, 212; lunar terminology in, 81; Pleiades terminology in, 113; solar terminology in, 67; symbolism of fireplace in, 176; terms for mother earth in, 175
Southern Cross: center of line of dark cloud constellations, 172; constellation of *ayllu* in Recuay, 134; heliacal rise and set dates of, 184; importance in celestial orientations, 59; Inca constellation, 130-131; Quechua constellation, 99, 102, 103, 104, 207, 208; in tropical forest astronomy, 147.
Spots on moon, 81, 213
Springs: as origin of rainbows, 88-89
Star: disappearance of, during day, 67; in female zodiacs, 79; general terminology, 106-107; identification of, in the field, 95; in Quechua cosmology, 199; mobile "star" of Collca, 121-122; single, and planets, 96-97
Star catalogue, 70, 96-105
Star map: drawn by S. Gaposhkin, 11; male and female understanding of, 79-80
Star-to-star constellations: 9; catalogue of, 97-102; general description and classification of, 107-109
Stein Callenfels, P. V. van, 109
Storehouse. *See Collca*
Stores: in Misminay, 21
Subterranean: arcs of rainbows, 88; hibernation of snakes, 179; hibernation of toads, 180-181; passage of sun, 64, 68; springs, source of serpents, 93; tunnel for sun, 69; water, and fecundity, 175-176
Sullaje (spray of light), 92
Sullivan, William, 208, 209, 211-212
Sulluullucu (constellation), 104, 178
Sun, 67-77; and agriculture, 27; birth of, in Inca cosmology, 201; haloes around, 90-91; in Inca cosmology, 196-197; Inca interest in, 199; and Milky Way, 61, 195; passage underground, 64; planting, 76-77; related to topography, 51; solstices, 70-71; three sections of, 71-74; twilight and, 151-152
Surucucu: serpent, 178
Suyus: crosses of the four, 136, 143-147; division of the sky into four, 55, 61; in Inca Cuzco, 42, 201; *pachapacariq ch'askas* of the, 159-166
Symbolism: of ancestors, 51; of animals, 191; of crosses, 147-150; of the earth, 174-176; of the four quarters in Misminay, 53; of Inca drums, 90; lunar, 79-81; menstrual, 85; of rainbows, 89; solar, 67-68; of sounds, 30-32; of urine, 176
Syncretism: astronomical, 4, 194; of calendars, 77; iconographic,

137-138; in solar terminology, 67-68; of solstices and saints' days, 70
Synodic lunar cycle: definition of, 23, 79; in the Inca calendar, 197; related to Milky Way, 165; terminology and conception of, in Quechua astronomy, 82-85; the three phases of, 214

Tachymenis peruviana, 177; birthing period, 179
Tahuantinsuyu, 47
Tastevin, P. C., 109, 178
Taurus, 98, 107
Taytacha: term for Jesus Christ and the sun, 67, 80
Taytapata: site of community cross, 51, 129
Tejeíro, Antonio, 5, 113
Teniente gobernador, 18
Terminology: lunar, 80-82; of celestial crosses, 134-143; of crosses, 131; solar, 67; of twilight and zenith stars, 155
Thief: divination of movement of, 64; lightning as, 91-92; rainbow as, 89, 90; shooting star as, 96, 208
Thom, A., 115
Thompson, Donald, 34
Thompson, Philip D., 213
T'ihsu: tilted axis of Milky Way, 61
Time: and the division of labor, 19; reckoning by sun, 68; and the senses, 32
Tinamou: breeding and incubation behavior, 184-185; cross of, 134; in the line of dark cloud constellations, 176; in the Milky Way, 55. *See also Yutu*
Tinkuy. *See* Union
Titicaca, Lake, 202
Toad: dark cloud constellation, 180-181, 208; mouth of, constellation, 98; race between tinamou and, 183
Torito (constellation), 97
Toro (star), 208
Trago: consumption during planting, 26; served in shells, 31
Traylor, Melvin A., 183
Trumpet: *pututu*, 30-31
Tuan, Yi-Fu, 54
Tuero Villa, Juan V., 11, 125-126, 193, 213
Turton, David, 7
Twilight: defined as periods of heliacal rise and set, 159; stars of the, 151-167; transitional time, 18

Ukhu: internal world, 38, 42, 63
Ullucu, 25
Uña Quilla (new moon), 81, 82
Uñallamacha (dark cloud constellation), 81, 104, 169, 170-171, 200
Underground. *See* Subterranean
Union: of celestial rivers, 60, 202; of opposites, 139; principle of, in cosmology, 150
Unuchillay (Inca constellation), 130
Ura moiety, 45
Uray moiety, 42
Urbano, Henrique-Oswaldo, 193
Urine: prohibition on urinating, 89; symbolism of, 176
Urkuchinantin: androgynous rainbows, 89
Urpi (dark cloud constellation), 104
Urqu (spots on the moon), 81
Ursa Major, 207, 213, 214
Urteaga, Horacio H., 6, 130
Urton, Gary, 10, 54, 62, 82, 130, 183, 187, 201, 211
Urubamba (Department of Cuzco): market in, 21; rise of Pleiades from, 114; subhorizon city from Misminay, 51
Urubamba Valley, 31, 64, 177. *See*

also Vilcanota Valley

Vagina: entry of rainbow through, 89
Valcárcel, Luis E., 87
Vallée, Lionel, 30, 193
Vargas C., César, 32
Vela, 100
Velasco N., J., 187
Venus: in Assyrian astronomy, 213; in Quechua astronomy, 35, 95, 97, 156, 166-167, 207
Vicuña: constellation, 208; trampling fox, 188
Vilcanota Valley: axis through center of the earth, 38; equated with Milky Way, 56, 64; in Inca cosmology, 201-204; in Quechua cosmology, 69, 71, 172; path of sun at night, 68. *See also* Urubamba Valley
Viracocha: creator god, 201; related to Milky Way and Vilcanota River, 204; symbol, 202, 215
Virgin Mary: constellation *(señoracha)*, 208; mother of the sun and Huayna Capac, 67; related to moon, 80-81
Virgins of the Sun, 79
Volcán (cosmic mountains), 37
von Dechend, Hertha, 212, 213
von Humboldt, Alexander, 173

Wachtel, Nathan, 4
Wagner, Catherine Allen, 19, 21, 26, 74, 175, 193, 212. *See also* Allen, Catherine
Wajran (star), 208
Wajus Qana (constellation), 208
Wakaya (constellation), 208
Walker, E. P., 189
Walker, W. F., 177
Wallis, Christopher, 11, 55, 90, 91, 107, 208, 214
Waman (constellation), 208
Wamanis (sacred mountains), source of water, 137
Wankar K'uychi (type of rainbow), 89, 90
Wañu (phases of moon), 82-85, 157
Wañumarka: cross on summit of, 129; and origin concepts, 48, 64; and foxes, 70, 189; principal sacred mountain of Misminay, 13; quarter of the north, 53; symbolism of, 48-52. *See also* Sacred Mountains
Warmi Mayor (moon), 80-81
Water: cross as symbol of, 138; and dark cloud constellations, 172-176; defining social space, 53; drunk by the sun, 68; haloes and, 91; orientation of flow of, 195; source of fertility, 93; terrestrial and celestial, 60; urine and subterranean, 89; *wamanis* as the source of, 137
Weaving: seasonality of, 30
Weather predictions, 11. *See also* Crop predictions; Divination; Prognostication
Weckmann, Luis, 138
Week, 21-23
Weiss, Gerald, 109
Wheat, 25; in crop rotation cycle, 32-34
Wheatley, Paul, 54
Wilbert, Johannes, 59, 147
Writing: Inca, 4

Yana Phuyu: 109, 169-191, 208. *See also* Dark cloud constellations
Yanacona, 13
Yucay (province of Urubamba): location and description of, 212; nature of calendar system in, 12; planting by moon in, 85, 122, 125; terminology for Scorpio in, 113
Yutu (dark cloud constellation), 104,

170-171, 181-185, 200, 207, 208. *See also* Tinamou; Urpi
Yutucruz (constellation), 102, 134-135, 208. *See also* Cross

Zenith: crossing of axes of Milky Way in, 61, 64, 65, 136, 196; defined as center sun, 75; in Inca astronomy, 197; midnight and noon stars in, 166-167; midnight star in, 96, 97, 107; movement of sun south of, 74; orientation by horizon and, 118; passage of sun in Misminay, 72; Scorpio in, in November, 178; stars and quarters of the moon, 157-158; in stellar divinations, 93; sun and the horizon calendar, 76; sun and inferior culmination of *Llamacñawin*, 187-188; terminology for midnight stars in, 153-156
Ziqpu Star: in Babylonian astronomy, 158
Zodiac: female lunar, 79, 198
Zuidema, R. Tom: xvii, 6, 7-8, 10, 21, 23, 34, 42, 43, 54, 64, 76, 82, 90, 130, 132, 134, 187, 196, 197, 199, 201